SCIENCE ACTIVITIES

Energy and Matter

Paul Spychal

Hodder & Stoughton

LONDON SYDNEY AUCKLAND TORONTO

About the book

This book arose as a result of the author's concern at the dearth of material suitable for teaching balanced GCSE science. The text is presented as an integrated series of activities each of which incorporates a main body together with a follow up section. Group discussion work is emphasised throughout. The questions attempt to develop student skills by encouraging them to think about science and its relevance to their everyday lives.

Dedication: For Peter, Carl and Adam

Acknowledgements

The publishers would like to thank the following for their permission to reproduce copyright material:

The Electricity Council for the Electrical Development Association advertisement 'Just the Difference'; The New Scientist for 'Irish advisers exploit African bogs' *New Scientist* 6 January 1983; Today for 'Experts comb horror house for new clues' *Today*, Monday 16 February 1987, 'Ribs clue to corpse' *Today* Friday 5 September 1986, 'Fly swarm uncovers a murder' *Today* 1 September 1986, and 'Baby mix-up mother still in nightmare' *Today* 1 September 1986 reproduced by kind permission of *Today* ©.

The author and publishers would like to thank the following for permission to use their photographs:

New Zealand High Commission London p5; Dr. F. Espenak/Science Photo Library p5; Oxfam p8; The Irish Peat Board p12; Barnaby's Picture Library pp13,49; Zanussi p14; Scottish Tourist Board p16; Ronald Sheridan's Ancient Art and Architecture Collection p17; Rex Features p20; Biofotos p21; Camera Press Ltd p23; BBC Hulton Picture Library p35; Redken Laboratories Ltd. p41; Ashmolean Museum Oxford p44; Science Museum London p57 (top and centre); The Royal Photographic Society p57 (bottom); John Walmsley p61.

First published in Great Britain 1990

British Library Cataloguing in Publication Data
Spychal, Paul
  Science activities: energy and matter
  1. Physical science. For schools
  I. Title
  500.2
ISBN 0 340 41484 7

Typeset in Ehrhardt by Gecko Ltd, Bicester, Oxon
Printed and bound in Hong Kong for the educational publishing division of Hodder and Stoughton Limited, Mill Road, Dunton Green, Sevenoaks, Kent by Colorcraft Ltd

Contents

List of Attainment Targets

1	Exploration of Science	10	Forces
2	The Variety of Life	11	Electricity and Magnetism
3	Processes of Life	12	IT and Microelectronics
4	Genetics and Evolution	13	Energy
5	Human Influences on the Earth	14	Sound and Music
6	Types and Uses of Materials	15	Using Light and e/m Radiation
7	Making New Materials	16	The Earth in Space
8	Explaining how Materials Behave	17	The Nature of Science
9	Earth and Atmosphere		

Science activist

This book makes you think about science. *Science Activities* tries to develop your science skills. By using the book you will become a *science activist*. You will be asked to:

Discuss

Have a chat and have a natter
Talk about the problem posed.
Don't argue or upset the teacher
Just make sure your mind's not closed!

Find out

What on Earth is a peat bog?
How on Earth do I find out?
What is meant by the "mass media"?
I might try an encyclopaedia!

Be mathematical

One and one make two I think
Although I cannot prove it.
Graphs to plot and to interpret
That's the way to do it.

Devise experiments

Think, think, think, think.
Worry, worry, worry, worry.
Plan, plan, plan, plan.
Mustn't do this in a hurry!

Draw

Pencil, paper, steady hand
Neatly draw and neatly label.
Subject's big so draw to scale
Then you won't fall from the table!

Design

Designer jeans, designer loo
It seems to be the rage.
I bet I could design as well
Just give me a blank page!

Write

ABC please help me
To write this corny letter.
XYZ then to bed
Could you have done it better?

Understand passages

What's this all about then Flo?
I haven't got a clue.
If you give it one more go
You'll know just what to do!

Sunshine

Hot stuff, geothermal power

Lights out for a total eclipse

Have you ever wondered where all our energy comes from? If you think the answer is the Sun you would be nearly correct. Except for nuclear energy and geothermal energy, all our energy does come from the Sun. Geothermal energy comes from the Earth's hot interior.

The Sun is our nearest star. It is at the centre of our Solar System. Earlier civilizations, like the Incas of Peru, worshipped the Sun. The Sun was their god. Can you imagine the Sun being switched off? What would happen to life on Earth?

At the centre of the Sun is its **core**. This core takes up about one tenth of the Sun's radius. Scientists think the temperature in the core is about 15 000 000 K. The visible surface of the Sun is called the **photosphere**. The temperature at the surface is only 6000 K! **Sunspots** are darker areas which appear on the surface in cycles of about eleven years. Sunspots are a little cooler than the surrounding photosphere. They give rise to bright eruptions called **flares** on the Sun's surface. These flares kick out charged particles which can reach Earth one day later. When the charged particles arrive at Earth they electrify the upper atmosphere. This can cause interference with radio communications.

During solar eclipses a bright red region called the **chromosphere** has been observed. It is a region, less than 100 km thick, of not very dense gas. The chromosphere gives rise to **prominences**. These are glowing gas jets ejected from the Sun to heights of over 300 000 km. Beyond the chromosphere is the **corona**. The corona, like the chromosphere, is visible only during a solar eclipse.

How can the Sun be so hot in the core? The answer could be because of **nuclear fusion**. This is when two heavy hydrogen nucleii are rammed together to form a larger helium nucleus, as shown in fig. 1. During fusion a little of the mass of heavy hydrogen is converted to energy. Every

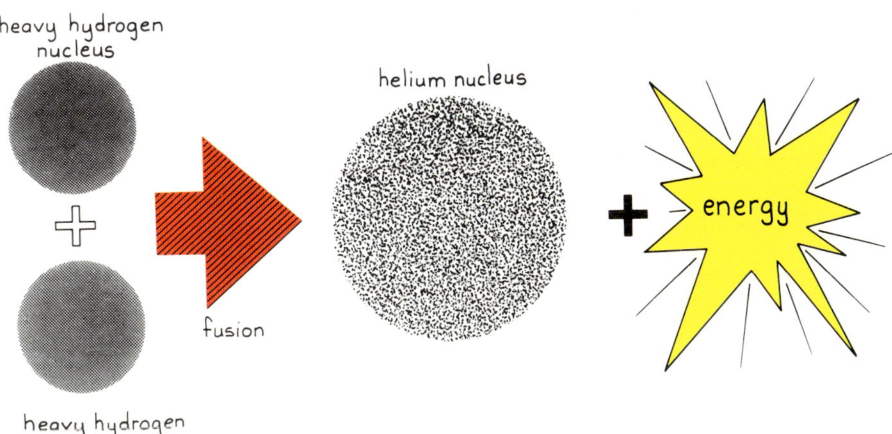

Fig. 1

Fusion

second the Sun loses over 4 million tonnes of mass which is converted to energy. This energy makes the core hot. It is the Earth's major energy source and keeps you alive. What happens to this energy when it arrives at Earth?

- Nearly 30 per cent of it is reflected back into space.

- About 47 per cent of it heats up the land's surface, atmosphere and oceans.

- Nearly 23 per cent of it is used to evaporate the water which falls back again as rain.

- About 0.2 per cent of it heats parts of the atmosphere and the oceans to different temperatures, causing winds, currents and waves.

- Only 0.02 per cent of it is used in photosynthesis by plants.

1 **a** Write down two energy sources on Earth that cannot be traced directly back to the Sun.
b What is the temperature in the Sun's core?
c Explain how sunspots can cause interference on your radio.
d How much of the Sun's energy arriving at Earth stays here?
e How much of the Sun's energy arriving at Earth is used by plants in photosynthesis?

2 Look at fig. 2. Copy and finish labelling the diagram of the Sun. Choose the correct labels from the box.

chromosphere   sunspot   core

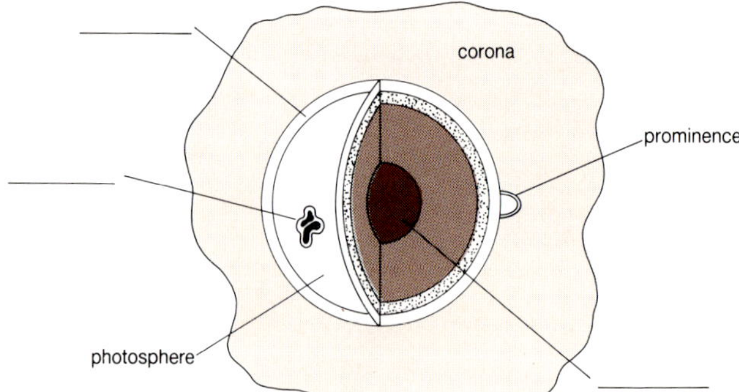

Fig. 2

Inside the sun

3 **a** Fig. 3 gives information about the Sun's composition. Put the information into a bar chart.
b In the passage is information about what happens to the Sun's energy when it arrives at Earth. Put this information in the form of a table. Your headings could be:

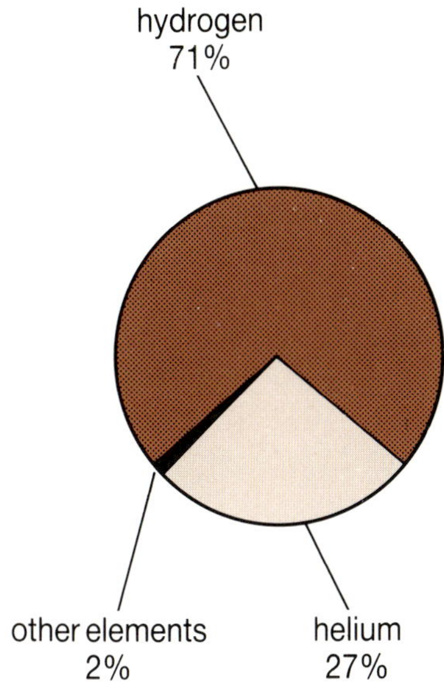

Fig. 3

The Sun's composition

- percentage of energy

- what happens to energy.

4 a Write down the Sun's surface temperature in K.

b Calculate the Sun's surface temperature in °C.

$$°C = K - 273$$

c The Sun's radius is 109 times larger than the Earth's. The Earth's radius is 6400 km. Calculate the radius of the Sun.

d The Sun's density is 0.255 times that of the Earth. The density of the Earth is 5500 kg/m^3. Calculate the density of the Sun.

5 Write a short article to explain what happens in nuclear fusion. A diagram will help your answer.

6 Answer this question in groups of four students. Much money is being spent on nuclear fusion research. Some scientists and governments hope that one day we may be able to copy, in fusion reactors, what happens in the Sun.

> The energy we receive from the Sun is about 25 000 times greater than the total world consumption of energy. That great fusion reactor in the sky already sends over one and a half million, million, million kilowatt-hours of energy every year. Why do we need to bother copying the fusion on the Sun?

Discuss the point of view shown in the box. Do you agree or disagree with it? Write a summary of your group's ideas. Draw a cartoon poster to illustrate your summary.

Solar flexers

1 Coal, natural gas and oil are examples of fossil fuels. Find out how they were formed. Draw a flow chart to summarise their formation.

2 Tidal power, wind power and hydroelectric power can all be used to make electricity. Choose one and write a description of how it is used to produce electricity. Diagrams will help your answer.

3 Who were the Incas of Peru? What were their beliefs about the Sun? Find out and write about the Incas. What other civilizations worshipped the Sun?

4 a Draw a diagram to explain the difference between night and day.

b Design a clock that uses the shadow cast by the Sun to tell time. When will you get the longest shadow?

What's for breakfast?

Growing maize

1 Millions of people start their day with a bowl of cornflakes. The seeds of cultivated grasses are called cereals. Maize is a cereal which is grown in many countries. It comes in many varieties. You may buy it as sweetcorn. It forms the basis of cornflakes.

a List three cereals, other than maize, which are commonly grown.

b What is a cultivated grass?

c Suggest reasons why cereals are the most important food in the world.

2 The Amaizing Corn Company has been developing four new breakfast cereals. Figs. 4, 5 and 6 give information about the contents of the new products.

a Which product has the greatest fat content?

b Which has the least fat content?

c Write down the protein content of bran flakes.

d Which product has the greatest carbohydrate content?

e Calculate the average protein content of the breakfast cereals.

f Why is there a large variation in protein content of the cereals?

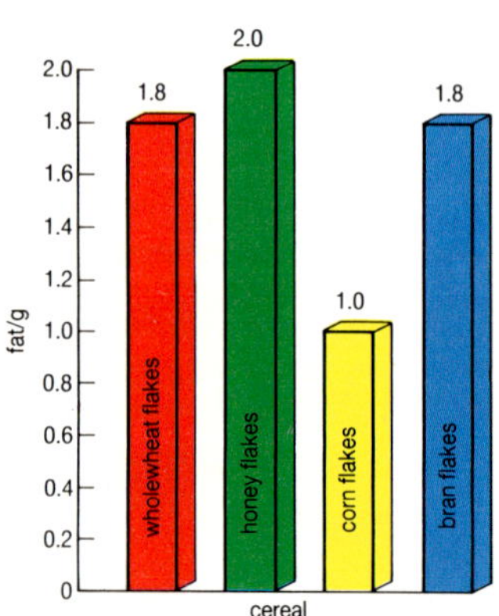

Fig. 4

Fat per 100 g of cereal

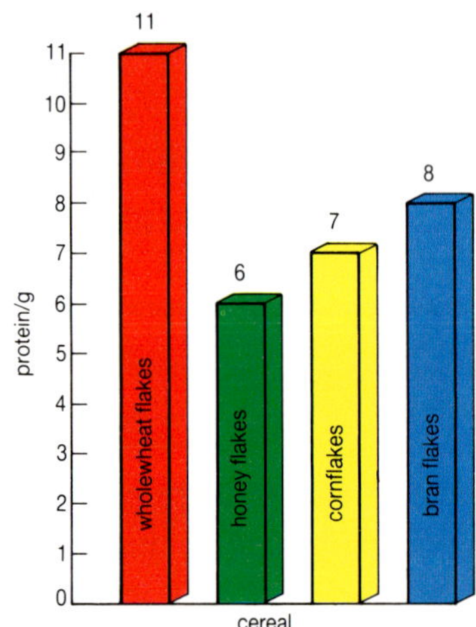

Fig. 5

Protein per 100 g of cereal

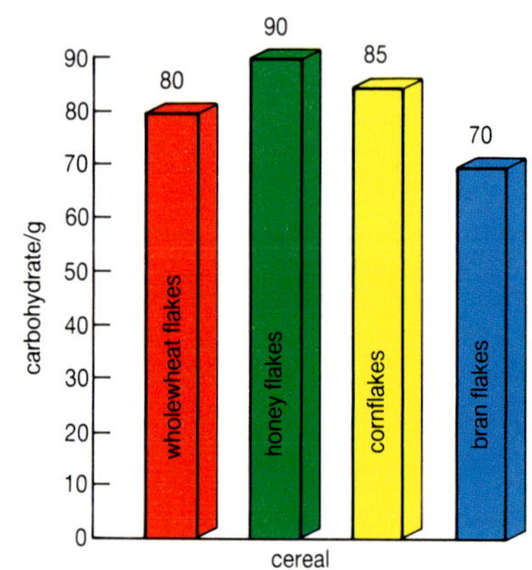

Fig.6

Carbohydrate per 100 g of cereal

breakfast cereal	fat/g	protein/g	carbohydrate/g
wholewheat flakes	1.8		
honey flakes		6.0	
cornflakes		7.0	
bran flakes			70.0

Table 1 Fat, protein and carbohydrate per 100 g of cereal

3 Copy and complete table 1. Use the information in figures 4, 5 and 6 to answer this question.

breakfast cereal	energy/kJ
wholewheat flakes	1500
honey flakes	1550
cornflakes	1490
bran flakes	1350

Table 2 Energy for breakfast per 100 g of cereal

4 a Look at table 2. Draw a bar chart to compare the energy content of the cereals.
b Calculate the average energy content of the new products.
c An average breakfast helping of cereal is 30 g. Calculate the energy content in 30 g of:

● wholewheat flakes

● bran flakes.

5 Answer this question in groups of three students. Imagine you are scientists working for the Amaizing Corn Company. The firm prefers its cornflakes to be crisp. However, the new cornflakes become soggy after the packet has been opened. You think this may be due to the moisture in the atmosphere. The Company has asked you to discuss and devise a simple experiment. You are to find out how the mass of cornflakes changes over several days. The cornflakes must be exposed to the atmosphere in your investigation.
a How do you expect the mass of cornflakes to change? Explain your answer.
b What equipment will you use?
c Briefly describe how you would carry out your investigation.
d Are soggy cornflakes tastier than crisp ones? How would you find out?

More corn

1 Amaizing wants to produce an eye-catching packaging design for its new cornflakes. Details of a 500 g packet of cereal are given in fig. 7. Design the new cornflakes packaging for Amaizing. Don't forget:

● there are six faces on the packet;

● you must give information about the contents to the consumer;

● advertising and competitions sometimes appear on the packet.

2 Do cornflakes get soggy more quickly in hot rooms than in cold rooms? Design a simple experiment to find out. List the measurements you would take.

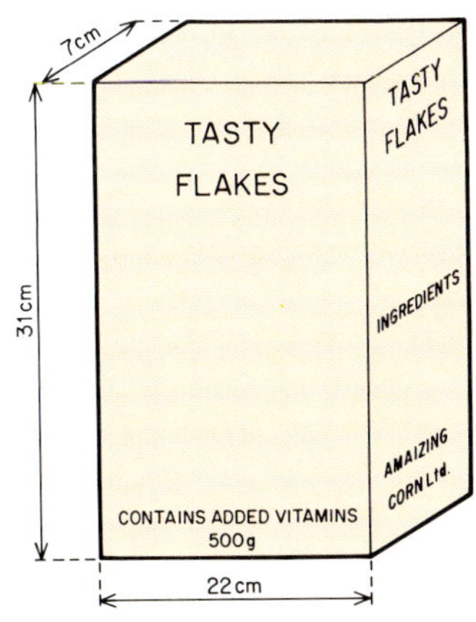

Fig. 7

Designer cornflakes

Feeding baby

1   Table 1 compares full cream cows' milk, human milk and powdered baby feed milk. Milk also contains minerals and vitamins not shown in the table.

 a   Which milk has the greatest energy? Which has the least?

 b   Which has the highest protein content? What must be done to the protein in cows' milk before it is suitable for babies to drink?

 c   Calculate the protein content in 1 litre of cows' milk.

 d   Draw a bar chart or charts to display the information in table 1.

per 100 mL of milk	cows' milk	human milk	powdered milk
energy/kJ	281	294	290
carbohydrate/g	5.0	7.3	7.2
fat/g	3.7	4.0	3.6
protein/g	3.5	1.1	1.5

Table 1 Comparing milk

% mums	baby's age/months
70	0
34	4
12	8
4	12
1	16

Table 2 Percentage of mums breast feeding

2   Many mums choose to breast feed their babies. Information about the number of mums breast feeding is given in table 2.

 a   How many mums breast feed their newly born babies? What happens to the number of mums breast feeding as their babies get older?

 b   Plot a graph of the information in table 2. Copy the grid in fig. 8 to start with.

 c   From your graph find out how many mums breast feed six-month-old babies.

 d   Suggest reasons why so many women have stopped breast feeding by the time the baby is six months old.

3   Babies need energy to grow and remain healthy. Table 3 gives information about a baby's energy needs.

 a   Does the baby need more or less energy as she grows older?

 b   Plot a graph to show how the baby's energy needs change with age. Copy the grid in fig. 9 to start.

 c   From your graph find out how much energy a six-month-old baby requires.

4   The DHSS recommends breast feeding. It is also suggested that mums wishing to breast feed should put their babies to the breast as soon as possible after birth. What conditions are necessary for a mum, immediately after giving birth, to get to know her baby?

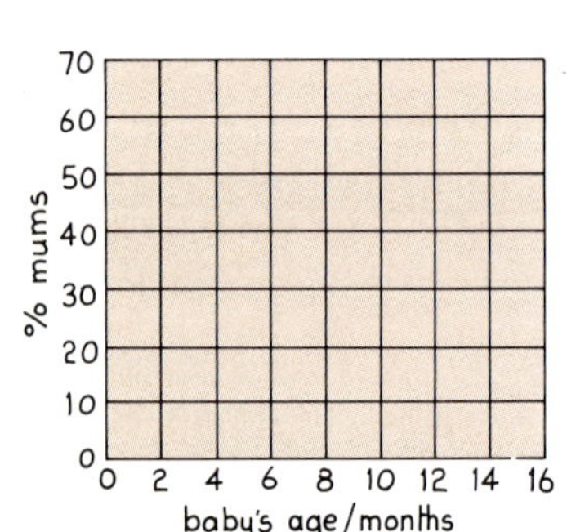

Fig. 8

Mums breast feeding

daily energy/kJ	age/months
2000	0
2750	4
3400	8
4000	12
4250	16

Table 3 Energy for growing

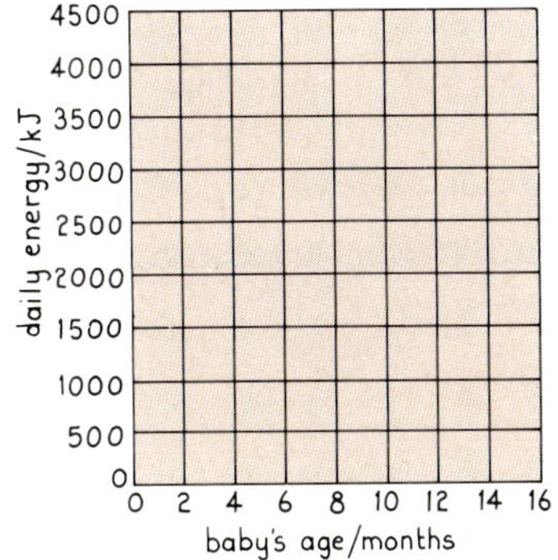

Fig. 9

Another graph

Fig. 10

Bottle feeding

5 Breast milk is natural, clean and protects the baby against infection. Babies absorb minerals such as zinc and iron better from human milk than from a powdered milk mix. They are also less likely to suffer allergic reactions to breast milk. Mums who start to breast feed may change to bottle feeding. Mums who start bottle feeding cannot later change to breast feeding.

a Suggest reasons why some women prefer to bottle feed.

b Does a mum who is breast feeding need more energy than one who is bottle feeding? Explain your answer.

c A balanced diet is necessary to ensure a good milk supply. Suggest foods which could be included in such a diet.

6 Care must be taken when bottle feeding. Babies are bottle fed with a specially modified cows' milk formula which has added nutrients. Instructions for bottle feeding might include:

- wash bottles and teats in detergent water then rinse

- sterilise teats and bottles by boiling in water or standing in sterilising solution

- empty out the solution

- fill the bottle up to the mark with recently boiled water

- add levelled scoops of powder as recommended by manufacturer

- screw on the cap and shake well.

a Why should the bottles and teats be sterilised?

b What might happen if too much powder is added to the bottle? Suggest how you might ensure a *level* scoop of milk powder.

c Why should the milk mixture be well shaken?

d Doctors suggest that a baby should not be left to sleep with a bottle in its mouth. Why?

Keep feeding

1 Draw a series of cartoons explaining how to change a nappy! Why do dirty nappies have to be sterilised before they are used again?

2 Write a short article comparing breast and bottle feeding.

3 Mums who have just given birth are often advised to get themselves fit by exercising. Why is this advice given?

4 When bottle feeding, teats need replacing every few weeks. If the hole in the teat is too small the baby may be underfed. When the teat hole is the right size the milk should drip at about one drop every second. Design a simple experiment to check the hole in a teat. Draw a flow diagram to show the stages in your experiment. List the equipment you need.

Peat bogs

An Irish peat bog

Have you ever been stuck in a peat bog? It's no fun cleaning your boots afterwards! Peat bogs cover vast areas in the temperate regions of the world. Peat is brownish black in colour and is fibrous and spongy. It can be burnt to provide heat. In parts of the world without coal, peat can be a good substitute. Unfortunately, when burnt it gives off lots of smelly smoke. Peat is a partly decomposed mass of vegetation. *Sphagnum* and *Hypnum* are mosses which make up peat. They need moist conditions to grow so they're quite happy on the surface of a peat bog. Underneath the surface previous generations of moss rot, without oxygen, and become compressed. Bogs may be tens of metres thick. Hang around and observe a peat bog for several million years! You might find, that given the right conditions, it turns into coal.

Carefully read the article on African bogs.

Irish advisers exploit African bogs

A team of experts from the Republic of Ireland is helping the small country of Burundi, in the heart of Africa, to set up an alternative fuel industry. The fuel is peat, and the centre of operations are Burundi's peat bogs. The experts, who work for the Irish Peat Board, are now as familiar with Burundi's bogs as they are with Irish ones.

Peat has always been a big source of fuel in Ireland. During peak production in the 1960s, the fuel generated over a third of the country's electricity. But up to now, Burundi has done little to exploit this resource, although it estimates its reserves as 500 million tonnes, as large as Ireland's. This makes Burundi potentially the world's sixth largest peat power, so the country's leader, President Jean Baptiste Bagaza, is keen to start exploiting it.

Bagaza sees peat as the main alternative to timber and imported diesel oil. It could cut fuel bills and, equally important, curtail the ravages of deforestation.

Continual cutting of wood for fuel has reduced Burundi's reserves to 90 000 hectares, just 3 per cent of the total land area. Burundi is already Africa's second most densely populated country, and because families have an average of eight children, the problem is getting worse. The inevitable consequence will be more trees cut down for fuel.

Bagaza hopes that exploiting peat will reverse the cycle. He wants to see women using peat charcoal in their small cooking stoves, and he hopes the country's young industries will also adopt this new fuel.

Over the past four years, over a dozen advisers from the Irish Peat Board have been seconded to a charity, Catholic Relief Services, which realised the potential for peat in Burundi. The US Agency for International Development has now joined the project, and is funding trial excavations at three Burundi bogs.

The three bogs yielded 14 000 tonnes of peat in 1982, most of which went to the army. The cutting technique is much the same as that used in Ireland. Labourers dig small trenches to drain the sites by gravity, dig out the peat, and dry it in the Sun.

The peat can either be burned in this state or carbonised to peat charcoal. According to Maurice Keane, a senior engineer with the peat board, peat has no calorific value until dried because in its natural state it has a moisture content of 90 per cent.

Unfortunately, as in the rest of Africa, Burundi's peat has high quantities of ash. This makes it difficult to ignite, and when it burns it gives off a great deal of smoke. So peat is unlikely to be a popular domestic fuel unless researchers can come up with a more suitable cooking stove.

The Irish-American project is funding work on a simple insulated stove, which will reach the ignition temperature more quickly than existing stoves. But whether or not Burundians will switch to different stoves is a different matter.

New Scientist   *6 January 1983*

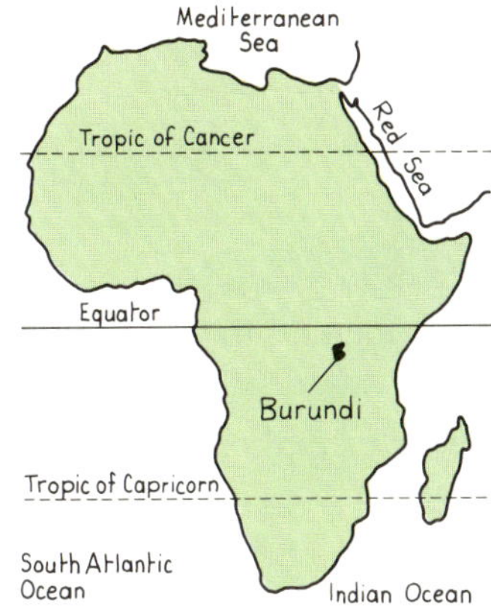

Fig. 11

Burundi

1 a What has happened to Burundi's trees?
  b Why does the President want his people to use peat for fuel?
  c Why should Irish scientists be helping Burundi?

2 a Why must peat be dried before being used as a fuel?
  b Explain, using a flow chart, how peat is removed from the ground.
  c What problem must be solved before the people of Burundi will accept peat as a domestic fuel?
  d Could peat be used to produce electricity in Burundi? Explain.

3 a What percentage of the total land area of Burundi is covered by trees?
  b Write down the area, in hectares, covered by trees.
  c Calculate the total land area of Burundi, in hectares.
  d Calculate Burundi's total land area in km^2.

$$100 \; hectares = 1 \; km^2$$

4 a What percentage of freshly dug peat is moisture?
  b How much peat was dug in Burundi in 1982?
  c Calculate the moisture content of peat dug in 1982.
  d Write down Burundi's reserves of peat.

5 Answer this question in groups of four students. Imagine you work for the Energy Minister in Burundi, fig. 11. Imagine also that a reliable, smokeless, peat burning stove has just been invented. Devise a strategy to convince the population not to burn valuable trees but to burn peat instead for fuel.
  a Write a letter to the Minister outlining your strategy.
  b Draw a poster describing your strategy.

More about moss

Neat meadow feather moss

Polytrichum – the coloured structures you can see are not flowers, but sporangia which contain spores for reproduction

1 Mosses are non-flowering plants. They helped to make the early world fit for animals to live in. There are several thousand species of mosses in the world. Find out and write about them.

2 Peat may be mechanically dried to form a fuel which can be used in engines. Write a short article describing the possible uses of peat.

3 Find out more about Burundi. Write about its people and industry. Draw a map showing Burundi and the surrounding countries. See fig. 11.

4 Plants may be grown to provide energy. Sugar cane, trees, seaweed and shrubs are examples of plants which could provide us with fuel. These plants are known as **biomass**. Find out about biomass. Write a short article explaining the possible uses of biomass.

Microwave cooking

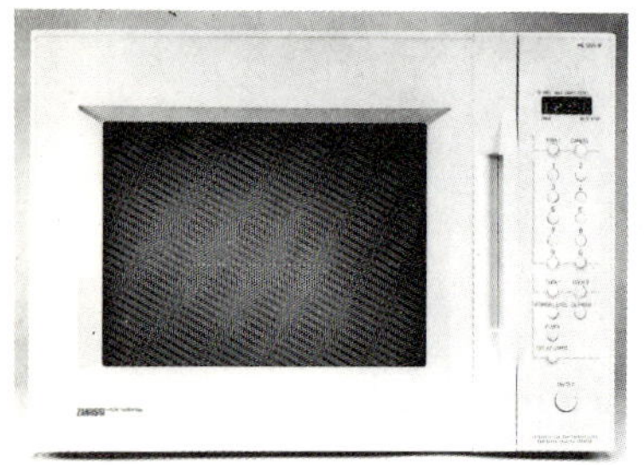

Hot stuff inside

Microwave cookers are becoming more popular in kitchens. Do you have a microwave cooker? How many of your fellow students have microwaves at home? Some salespersons will tell you that microwaves make perfect cookers. Should you believe them?

Microwaves are **electromagnetic waves** with a wavelength between 1 mm and 30 cm. Microwaves form part of the **electromagnetic spectrum**. Electromagnetic waves are caused by charged particles accelerating. These waves are transverse waves and don't need a medium to push against, unlike sound. All electromagnetic waves travel at the velocity of light which is 300 000 000 m/s. Gamma-rays, X-rays, ultra-violet radiation, visible light, microwaves and radio waves are all examples of electromagnetic waves.

Microwave cooking is said to be quick, cool, convenient, clean, versatile, economical and energy efficient! Food cooked in a microwave is supposed to keep its flavour, colour and nutritional value. This sounds almost too good to be true, there must be a catch somewhere! Look at fig. 12 which shows how the microwave oven works. A supply unit, inside the cooker, activates a **magnetron**. The magnetron produces the microwaves. The microwaves now travel along a **wave guide** to the oven. Here they may come in contact with a special metal stirrer. The stirrer sends waves to all areas of the oven.

When microwave energy meets a substance it can be reflected or absorbed or it can go straight through. Metals reflect microwaves. Paper, glass, china, pottery, plastic and wooden baskets allow microwaves straight through. The moisture in food absorbs microwave energy. The microwaves make the moisture molecules vibrate to and fro, about thirty thousand million times every second! These vibrations generate heat which cooks your food.

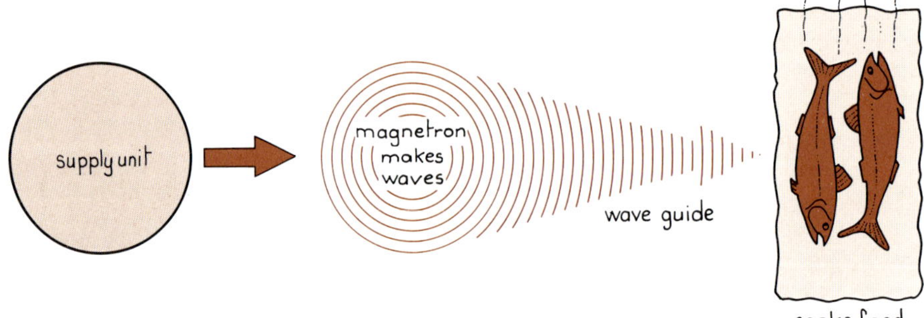

Fig. 12

Making waves in the kitchen

1 a   List six examples of electromagnetic waves.
  b   Write down three things all electromagnetic waves have in common.
  c   What are electromagnetic waves caused by?

2 Copy and complete the following sentences. You will find the answers in the passage above.

  a   All . . . . travel at the velocity of light.

  b   Moisture . . . . , in food, are vibrated by microwaves.

  c   Aluminium foil . . . . microwaves.

  d   Food . . . . microwaves.

  e   Inside a microwave cooker the waves are produced by a . . . . .

3 Look at fig. 13. It shows the power control settings on a typical 700 watt microwave cooker. On the high power setting all 700 watts are used. On the low power setting only 150 watts are used. The different power settings work by pulsing the microwave energy on and off for varying times.

  a   Now copy and complete table 1.

  b   Calculate the maximum current, in amps, the microwave cooker would use when a voltage of 240 volts a.c. is supplied.

$$power = current \times voltage$$

  c   What fuse would you use for the cooker?

4 Work in groups of four students to answer this question. Imagine you are sales directors for a large electrical company selling a new range of microwave cookers. You are to discuss and decide upon a set of guidelines for your company salespersons.

    Your guidelines should advise your employees on what to say to prospective microwave customers. Write down your company guidelines for salespersons. Use company headed paper of your own design.

Food for thought

1 Radar is short for 'radio direction and ranging'. It is a system which locates distant objects by means of reflecting radio waves. Radar was developed in Britain by Sir Robert Watson-Watts and a team of scientists. Find out about radar and what it is used for. Write an article on radar. Labelled diagrams may be useful in your article.

2 Design the inside of a microwave cooker. Don't forget that:

- the inside walls must be made of a material which reflects microwaves, otherwise the cooker will not be very efficient;

- the magnetron must stop generating microwaves when the door is open, for safety;

- when cooking, the microwaves must be 'spread out' so that the food cooks as evenly as possible;

- it helps to have the dimensions or measurements of a design.

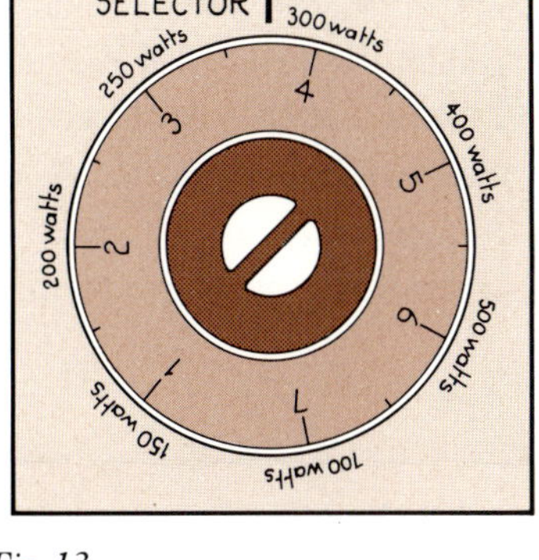

Fig. 13

Power control settings

setting	power/W	% power used
1		
2		
3		
4		
5		
6		
7	700	100

Table 1 Percentage of power used by cookers

Bagpipes and all that jazz

Bagpipes

Music has been part of human life for thousands of years. Ancient flutes and whistles dating back 25 000 years or more have been found on several sites. In ancient Greek the word music means the language of the Muses. The Muses were the nine Greek goddesses of the arts, daughters of Zeus.

Bagpipes may have been played by the ancient Hittites 3500 years ago. Emperor Nero, of Rome, is said to have enjoyed playing the bagpipes. The Romans are believed to have brought them to Britain! They are now played in many forms and in countless countries around the world. Bagpipes consist of:

- a leather bag to hold the air which works the pipes;

- a tube to blow up the leather bag, by mouth or small bellows;

- a chanter or pipe to play the melody;

- 2 or 3 drones, each making a fixed continuous sound.
  Bagpipes have a unique sound. They have even inspired poems and nursery rhymes.

1 a Explain the origin of the word *music*.
b Where might the word *museum* have originated?
c Suggest reasons why bagpipes are played in many countries.

2 Look at the photo. Describe how the player produces a note on his bagpipe. Can he control the loudness of the music he plays? How many notes can he play simultaneously?

3 Information about the diatonic scale of notes is given in fig. 14.
a How many notes are there in the octave shown?
b Do the notes on the scale increase by equal amounts?
c What frequency has the note *f* shown in fig. 14?
d Write down the frequency of the note *f* an octave higher.

Fig. 14

The diatonic scale

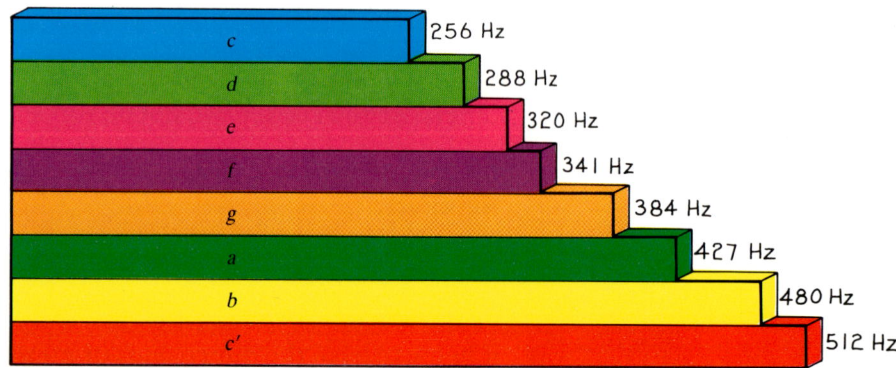

4 a Put the information given in fig. 14 into a table. Your headings might be *note* and *frequency*.
b Calculate the ratios:

- d/c, this is a major tone;

- e/d, this is a minor tone;

- f/e, this is a semitone.

c Explain the difference between major, minor and semitones.

5 Imagine you are prehistoric! You live in the Stone Age, 20 000 years ago. Your tribe are travelling hunters. Design a simple musical instrument which you could play. Be original!
a What materials might you use? Briefly describe how you will make your musical instrument.
b Sketch your design.
c Explain how you would play the instrument. Describe how it works.

More jazz

1 Look carefully at the photo. It shows a relief from the royal palace in Nineveh.
a What musical instruments can you see?
b What else do you observe in the photograph?
c Draw a diagram explaining how one of the instruments might have been played.

Music of Assyria 900 BC

2 Find out and write about two of the instruments in the box.

barrel-organ	harpsichord	hurdy-gurdy
lute	synthesiser	zither

Cool customers

Three students discussed a team project to find out about heat loss. Augusta, Beryl and Cecil wanted to understand how buildings and animals lost heat. They agreed to do an investigation in a draughty part of the laboratory. Some of the equipment they used is shown in fig. 15. They also used an electric kettle to supply hot water to the three identical cans. Augusta's can was uninsulated. Beryl used wet felt lagging around her can. Cecil's can was insulated with dry felt. The team agreed to take temperature readings every four minutes. They plotted their results on graph paper as they carried out their investigation. Their results are shown in fig. 16.

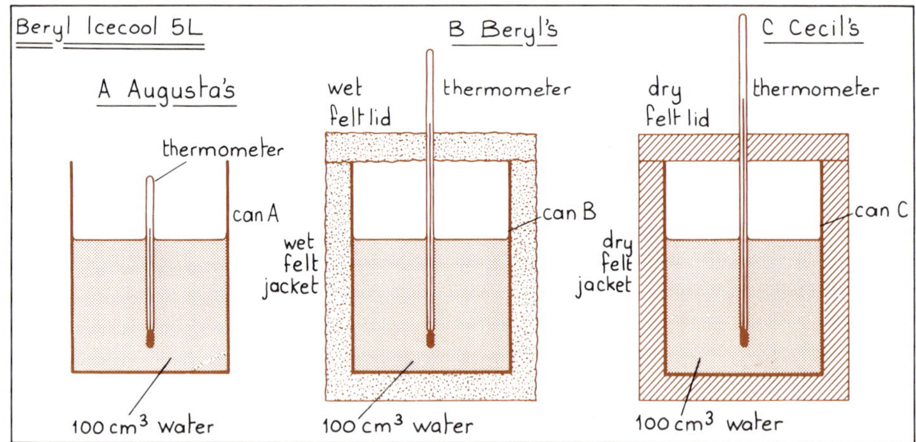

Fig. 15

A cool project

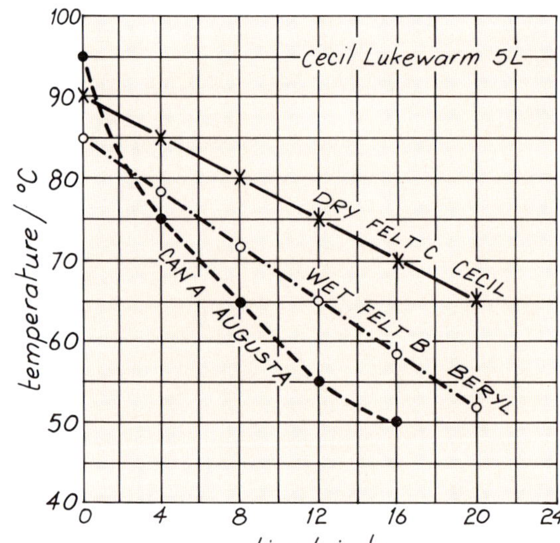

Fig. 16

S – cool results

1 **a** How much water did each can contain?
   b Write down the starting temperature of each can.
   c What can you say about the thermometers used by the students?

	temperature/°C		
time/min	A	B	C
0			
4			
8			
12			
16			
20			

Table 1 Project results

energy lost	% heat lost	
from	uninsulated	insulated
walls	35	12
roof	25	8
floor	15	10
draughts	15	12
windows	10	5

Table 2 Semi-detached heat loss

Fig. 17

Heat lost by you

d For how long did they carry out their investigation?

e How would you improve their investigation to make it fair?

f List all the equipment they needed to carry out their investigation.

2 Copy and complete table 1. Estimate your answers from fig. 16.

3 a Calculate the temperature drop of each can after sixteen minutes.

b What conclusions might the three students have come to? Write about the effect of wet and dry lagging in your answer.

c Beryl touched the metal outside of each can after two minutes. What do you think she noticed?

d Why did the students agree to carry out their investigation in a *draughty* part of the laboratory?

4 Look at table 2. It gives information about heat lost from a three bedroomed semi-detached house.

a Describe how a fully insulated house differs from one that is uninsulated.

b What percentage of energy can be saved by fully insulating a semi-detached?

c Draw bar charts to compare heat lost from an uninsulated house to heat lost from an insulated house.

d How do you expect the percentage heat lost to differ for a terraced house?

e Would a detached house lose more or less energy than a terraced house? Explain your answer.

f Look at fig. 15. Which two experiments might be used to compare heat lost from an uninsulated and an insulated house? Explain your answer.

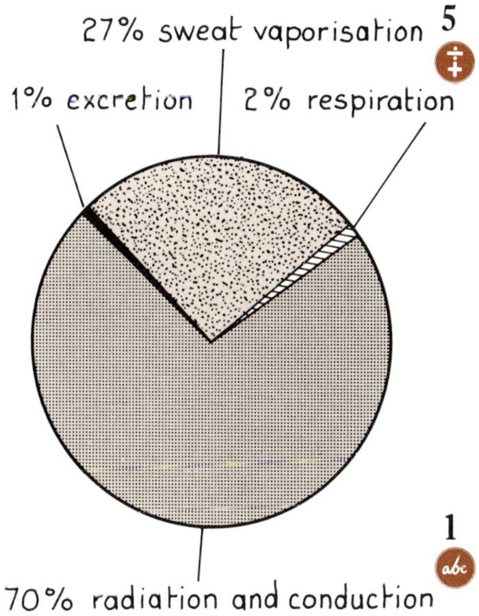

5 Fig. 17 gives information about how body heat is lost by a human.

a How much body heat is lost by sweating?

b How might the body lose heat by excretion?

c Put the information shown in fig. 17 into the form of a table.

d How do you think the human body produces heat?

e Humans have only a thin layer of hair. We keep warm by wearing clothing. How do birds and bears keep warm? Use diagrams to explain your answers.

Turning on the heat

1 **Hypothermia** can be caused by damp clothes and draughty conditions. Sometimes it leads to death.

a Find out and write about the causes and treatment of hypothermia.

b Look at fig. 15. Which two experiments might be used to compare heat lost by humans in wet and dry conditions? Explain.

c What clothing should you wear when walking in the hills?

Energy efficient seals

1 Copy and complete the passage on energy. Choose your answers from the box.

heat	energy	work	capacity
converts	electrical	devices	nuclear

Energy is the .... to do ..... Sound, heat, mechanical, light, chemical, ... and .... are all forms of energy. .... can convert one form of .... into one or more other forms of energy. An electric fire, for example, ... electrical energy mainly to .... energy.

2 Copy and complete table 1 on devices.

device	energy before	energy/ energies after
electric kettle	electrical	heat
light bulb		
loudspeaker		
cell		
dynamite		
microphone		

Table 1 Devices

Rocking around the world

3 The stages in the live transmission of a rock concert in the USA to the United Kingdom are:

- camera watches concert in New York, USA;
- pictures transmitted to local TV centre in USA;
- signal sent to ground control station in USA;
- USA transmitting dish receives signal;
- signal sent to communications satellite, over Atlantic;
- forwarded to receiving dish aerial in UK;
- UK ground control station receives signal;
- signal forwarded to UK TV broadcasting station;
- transmitting mast sends signal throughout UK;
- received by UK domestic receiving aerial;
- picture appears on television set!

a Write down the energy conversions that take place in the transmission. Start with the musicians playing their music.

b Telephone calls may be sent across the Atlantic in a similar way. However, the USA TV centre is replaced by a telephone exchange. Also the UK ground control station forwards the call to a telephone exchange. Draw a flow diagram to show how a telephone call from Washington, USA, reaches your home.

4 **a** Many energy changes take place in a car. Draw a series of cartoons to show the changes that occur. Start with a cartoon of the ignition being turned on.

b A petrol car is only about 25 per cent efficient. What happens to the other 75 per cent of energy?

5 Animals can transfer energy to their offspring. A dairy cow can convert 30 per cent of her daily energy output into milk. Grey seals, change 57 per cent of their energy output into milk for their pups. Seals seem to be able to turn their body fat into milk very efficiently. Unfortunately, seals are clumsy on land. This means they are easy targets for predators. They cannot, thus, spend much time on land giving milk to their pups. While she is on land feeding her pup the mother seal uses up 84 per cent of her energy reserves.

a Explain the first sentence in the passage.

b A mother seal spends only eighteen days on land feeding her pup. Why doesn't she stay longer?

c A mother seal loses about 65 kg of stored fat when feeding her pup. Her offspring gains 30 kg during this time. Explain the difference in the figures.

d The dairy cow converts only 30 per cent of her energy output into milk. What does she use the other 70 per cent for?

e How might scientists have obtained the information about grey seals?

6 Answer this question in groups of four students. Read *The Daily Nag's* scoop article, fig. 18. Do you think Jools Wot is correct in his claim? Write a short summary of your group's ideas. Design a simple experiment to find out if Jools is correct. Draw a flow diagram to describe your group's experiment.

Energy matters

1 James Prescott Joule was born in Salford in 1818 and died in 1889. He was a scientist who was fascinated by the conversion of mechanical into heat energy. Find out and write about him.

2 List six kitchen devices used in your home. Make an energy table, similar to table 1, to include your kitchen devices.

3 Choose either a reptile or a bird. For your chosen animal:

- find out what it eats

- find out how it uses up its energy in its daily activities

- draw a labelled diagram to show its main features.

4 Write a letter to *The Daily Nag* newspaper arguing for or against Jools Wot's claim.

The seal of approval

Fig. 18

Scoop for The Daily Nag

Ecstatic

Humans have known about the effects of static electricity for thousands of years. Thales was a Greek philosopher who lived about 2500 years ago. It is said that one day he stumbled across a piece of amber. After picking it up he rubbed it on his sleeve to clean it. But the amber accidentally fell and landed on a pile of dried leaves. He was suprised when he picked it up again. This was because bits of leaf had attached themselves to the amber. He called the effect electricity after *elektron*, the Greek word for amber.

Stephen Gray, in the early eighteenth century, carried out many electrical experiments. In one he suspended a young schoolboy by two silk ropes! Gray then applied a charge to the boy's feet. The poor lad was suprised when Gray drew sparks from his face. Clearly the young boy could not have been too ecstatic!

1   **a**   Who was Thales? When did he live?

     b   Explain the origin of the word *electricity*.

     c   List the stages in Gray's experiment.

     d   Stephen Gray and the boy may have come to different conclusions about the experiment! Write a few lines to suggest what each may have found out.

2   Imagine you are the schoolboy's mum. Write a brief letter to Stephen Gray complaining about your son's treatment.

3   Many strange theories to explain static electricity have been invented. Two such theories, used to explain Thales' observations, are shown in fig. 19.

     a   Write a short passage to explain the magic ray theory.

     b   Use a diagram to describe the condensing vapour theory.

4   Answer this question in groups of two students. Discuss and design a simple experiment. Your experiment will test one of the theories shown in fig. 19.

     a   List the equipment you need.

     b   Describe briefly how you would carry out your investigation. What result would prove the theory correct?

     c   Give a short talk to your class to explain your experimental design.

5   Joey carried out a simple investigation, fig. 20. She rubbed a plastic pen on her jumper sleeve. Joey then used the pen to pick up a piece of paper from the bench. The paper than fell off the pen and hit the bench. Immediately it jumped off and attached itself to the pen again.

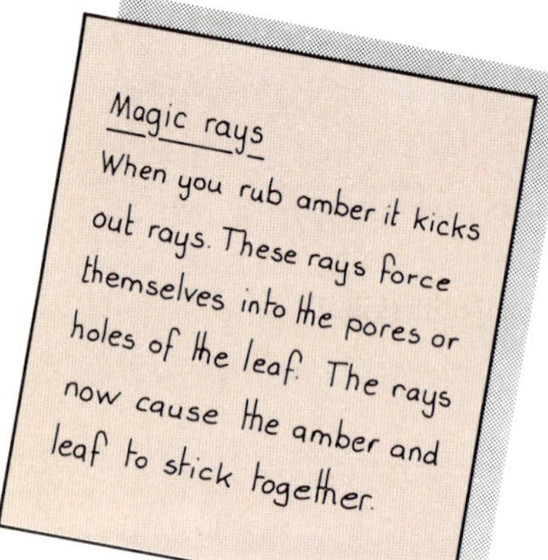

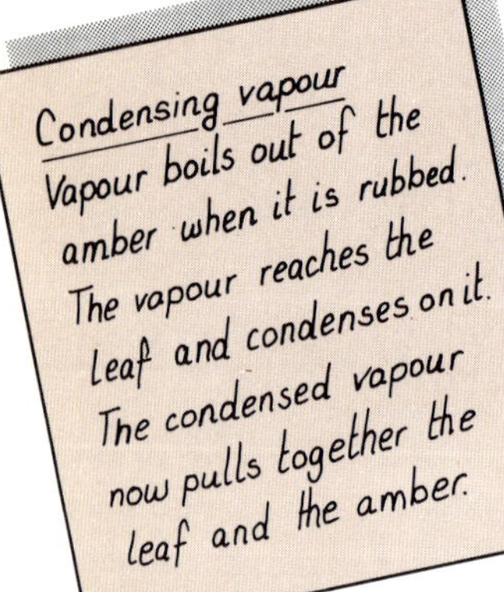

Fig. 19

Explaining statics

a Which part of the pen attracted the paper?

b Which part of the pen would not attract the paper? Explain your answer.

c Explain what happened in Joey's experiment. Don't forget that:

- matter consists of positive and negative charges

- like charges repel and unlike charges attract.

Fig. 20

Joey's experiment

Charge of the Light Brigade

1 Look at the nasty storm shown in the photo.

a What process does it show? Find out why it occurs.

b How might tall buildings be protected from the process?

c Write a short poem to describe your feelings during such a storm.

2 a List four everyday occurrences of static electricity.

b Briefly explain the difference between conductors and insulators.

c Imagine you have been given an unfamiliar material. Design a simple experiment to find out if it is an electrical conductor. List the stages in your experiment.

3 Find out and write about the following scientists:

- Sir William Gilbert, Queen Elizabeth I's doctor, who lived between 1544 and 1603

- Benjamin Franklin, who lived between 1706 and 1790, an American who helped draw up the Declaration of Independence.

4 a Explain the following uses of static charges:

- spraying pesticides on crops

- spraying paint on walls.

b Why might plastic materials and man made fibres be banned from hospital operating theatres?

Stormy weather

Important connections

1 Gullusan carried out an electrical investigation during a science lesson. Her rough notes are shown in fig. 21.

    a Sketch the circuit symbols for an ammeter and a voltmeter.

    b Which meter did she connect in series? Which meter is in parallel?

    c How did Gullusan control the current in her circuit?

    d Describe briefly the energy changes that take place in the circuit.

    e Draw a circuit diagram of Gullusan's experimental arrangement.

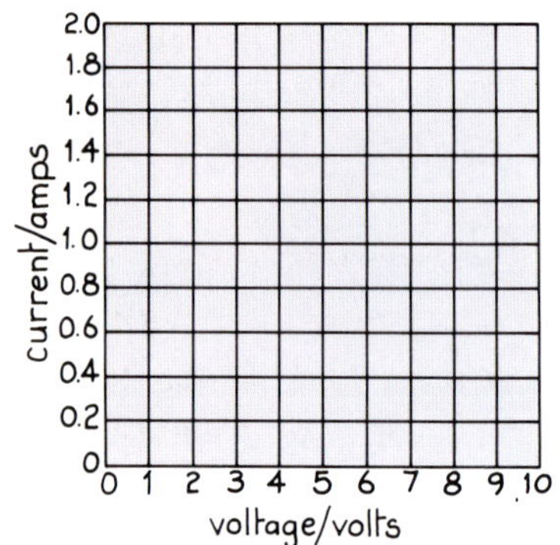

voltage /volts	current /amps	resistance /ohms
0	0	0
2	0.7	2.9
4	1.1	
6	1.4	
8	1.6	
10	1.8	

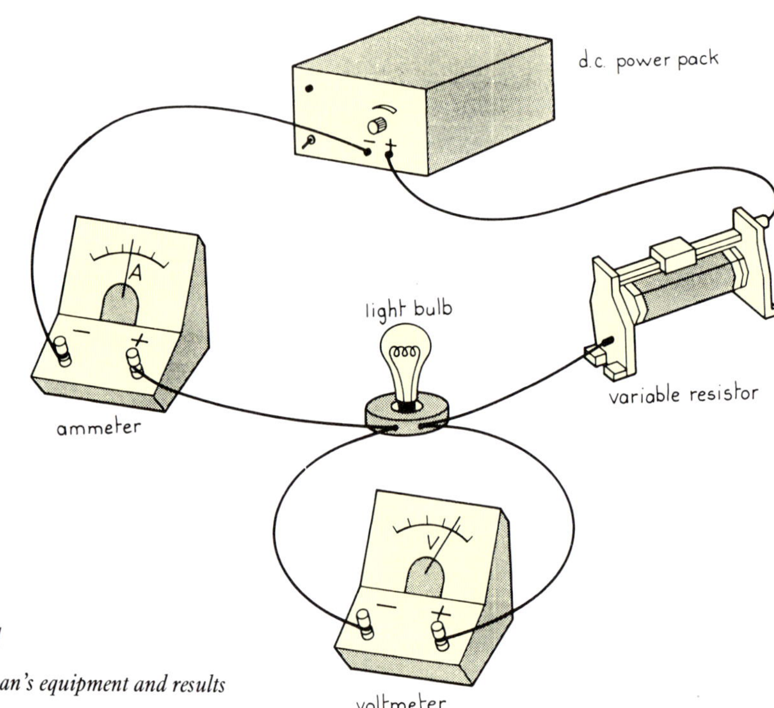

Fig. 21

Gullusan's equipment and results

2 a Copy and complete the graph in fig. 21. Use graph paper. Draw a smooth curve through the points.

    b Describe the shape of the graph.

    c Use your graph to predict the circuit current at 12 volts.

3 a Copy and complete Gullusan's results table.

$$resistance = voltage \div current$$

    b What happens to the resistance of the bulb as the circuit current increases? Suggest why this happens.

    c List the safety precautions that should have been taken by Gullusan.

    d Describe briefly how *you* might have carried out her investigation.

    e Why is tungsten used as the filament in light bulbs?

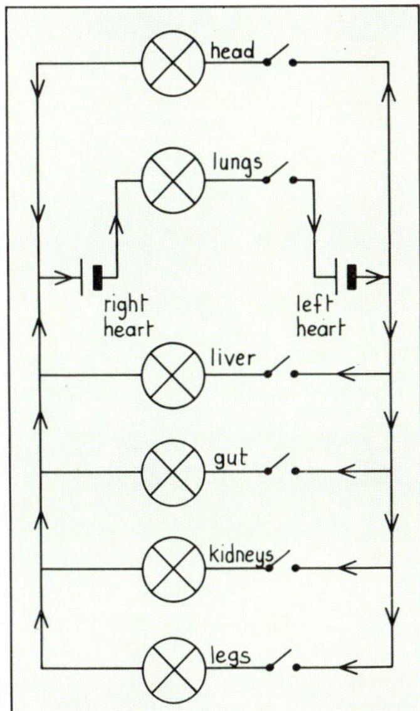

Fig. 22

Gullusan's model man

Fig. 23

Christmas lights

Fig. 24

The importance of kitchen layout

4 Gullusan designed an electrical model, fig. 22, of a man's circulation.

    a   What did she use to represent the two sides of his heart?

    b   How were the other organs of the body represented?

    c   What do the arrows mean in her model?

    d   Suggest why she placed the organs in parallel.

    e   Summarise the circulation system shown in fig. 22.

    f   The left side of the heart works harder than the right. Suggest why this is the case. Re-design Gullusan's model to account for this fact.

5 Answer this question in groups of three students. Fuse wire is normally marked with a rating. This is the maximum current, in amps, it can take before it melts. Imagine you have been given an unmarked reel of fuse wire. Discuss and design an experiment to find out its rating. Don't forget that fuse wire gets hot before it melts!

    a   List the equipment you need.

    b   Sketch a circuit diagram of your design.

    c   What measurements will you take?

    d   Briefly describe how you would carry out your experiment. Don't forget to write down your safety precautions.

Shining examples

1 Describe a Christmas tree light circuit. Use the equipment shown in fig. 23. Your design must be such that when one bulb blows the others stay on. Describe how your circuit works.

2 Wheelchair kitchen users may find conventional kitchens awkward to use, fig. 24. Kitchen layout is important to the disabled. Design and sketch a kitchen layout suitable for a wheelchair user. Take care when choosing the electrical appliances for your design. Suggest suitable positions for the sockets and light switches in the kitchen.

Power to the people

1 Look at fig. 25. It shows the fuels used in United Kingdom power stations.

    a List three fuels used in power stations.

    b What percentage of the fuel used in 1983/4 was oil?

    c How did nuclear fuel usage change between 1982 and 1985?

    d What happened to coal usage between 1982 and 1985?

    e Why did these *striking* changes occur?

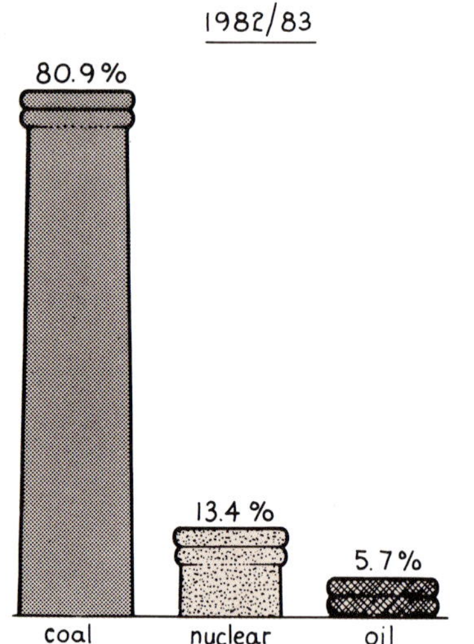

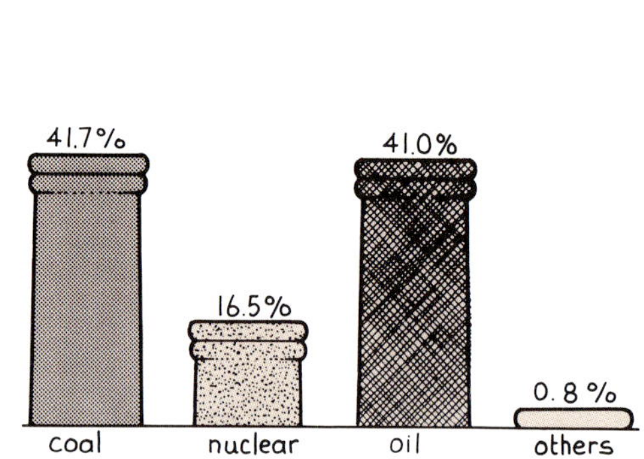

Fig. 25

Fuel used in United Kingdom power stations

2 **a** What happens to fossil fuels in power stations?

    b Write down one example of an *'other'* fuel shown in fig. 25.

    c Put the information given in fig. 25 into a table. Your table headings could be *year, type of fuel* and *percentage usage*.

3 Discuss this question in groups of four students. The fuel used in power stations may change by the year 2000. What will fig. 25 look like in AD 2000? Some points you might wish to consider in your discussion are:

- how long will fossil fuels last?

- how acceptable is nuclear power?

- are there other sources of energy?

Write a summary, including a bar chart, of your group's ideas.

4 Look at table 1. It gives information about United Kingdom electricity sales.

    a Write down the major consumer of electricity in 1920.

year	total sales/ thousands of GWh	domestic %	farming %	commercial %	industrial %	other %
1920	3.2	8.4	0	10.9	67.0	13.7
1930	8.2	17.2	0.1	14.5	57.1	11.1
1940	21.8	25.8	0.3	12.3	56.5	5.1
1950	38.3	32.6	0.9	11.9	50.1	4.5
1960	85.7	31.3	1.8	12.2	50.4	4.3
1970	168.2	37.5	1.7	15.0	42.1	3.7
1980	204.8	35.9	1.6	18.5	40.8	3.2
1985	203.0	36.1	1.5	21.7	37.5	3.2

Table 1 United Kingdom electricity sales; GWh = gigawatt hours, 1 GW = 1 000 000 000 W

b How have total sales of electricity changed between 1920 and 1985?

c Draw bar charts to show how sales to domestic and industrial users changed between 1920 and 1980.

d What changes in electricity sales might occur in the next twenty years?

Switch on

1 Erica and Chrysos did an experiment using a transformer. Their results are shown in fig. 26.

    a Did they use a step-up or step-down transformer?

    b Plot a graph of *input* against *output*.

    c What value of input would have given an output of 21.0 V?

2 Answer this question in groups of four students. Discuss the point of view shown in the box.

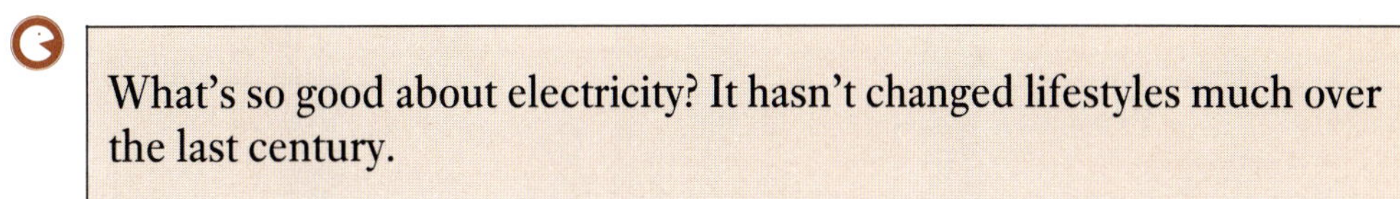

Briefly write down your group's ideas. Design a poster to compare life in the nineteenth and twentieth centuries.

3 Power stations need water. After use the water is still hot. How might you use this hot water to benefit your community? Use a labelled diagram to illustrate your answer.

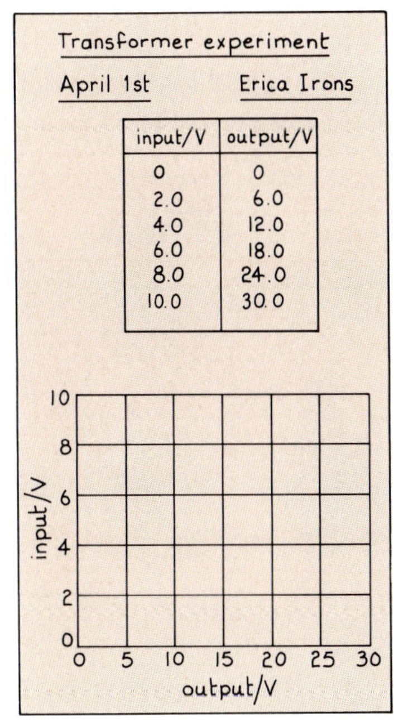

Fig. 26

Erica's results

Shocking figures

Electrical devices or appliances may help to cut down the work at home. How many electrical devices do you have at home? Which devices are the most useful to you, your mum and dad? Look at table 1 which relates to the United Kingdom. It shows how ownership levels of electrical appliances have changed between 1955 and 1985. Figures in the table relate to the percentage of domestic customers.

year	% cookers	% lawn mowers	% black and white televisions	% colour televisions	% video recorders	% washing machines
1955	25	0	35	0	0	18
1966	37	0	91	0	0	60
1978	44	25	51	60	0	73
1985	48	53	33	88	28	83

Table 1 Ownership of appliances, United Kingdom

1 a Which appliance, in 1955, was owned by the most people?
b In 1985 which appliance was owned by the least people?

2 a Draw two bar charts to compare how ownership of *colour* and *black and white* televisions changed between 1955 and 1985.
b Why have these changes occurred?

3 Draw graphs to show how ownership levels of the appliances shown in table 1 changed between 1955 and 1985.

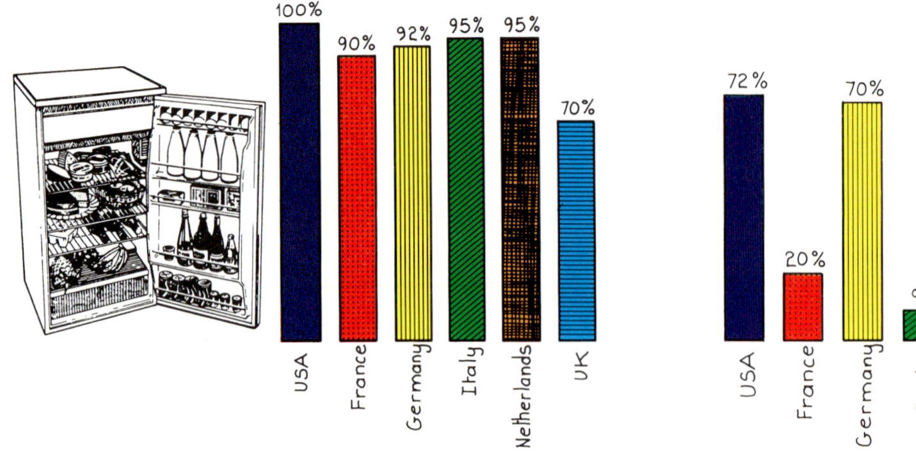

Fig. 27

Refrigerator ownership

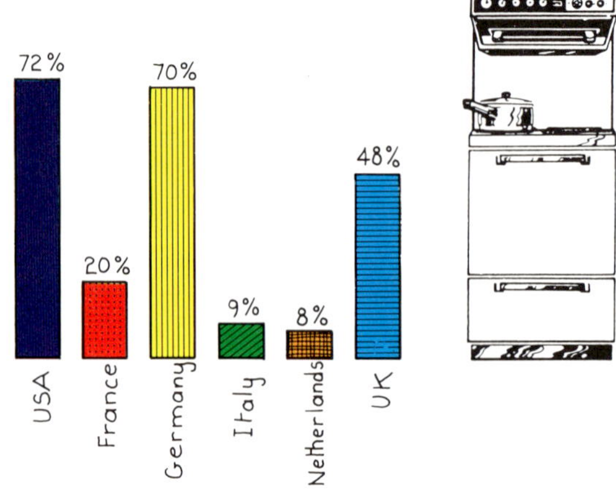

Fig. 28

Electric cooker ownership

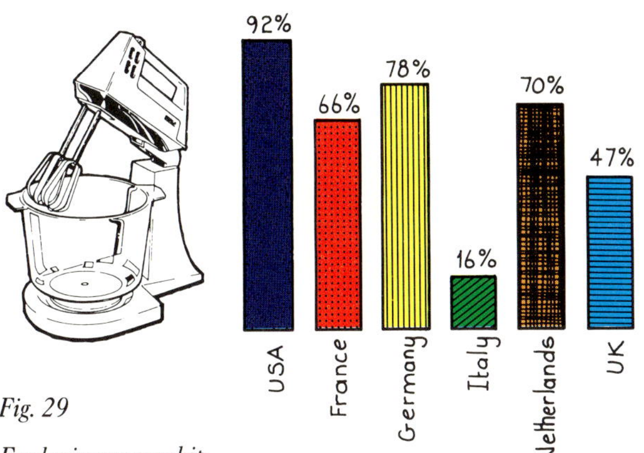

Fig. 29

Food mixer ownership

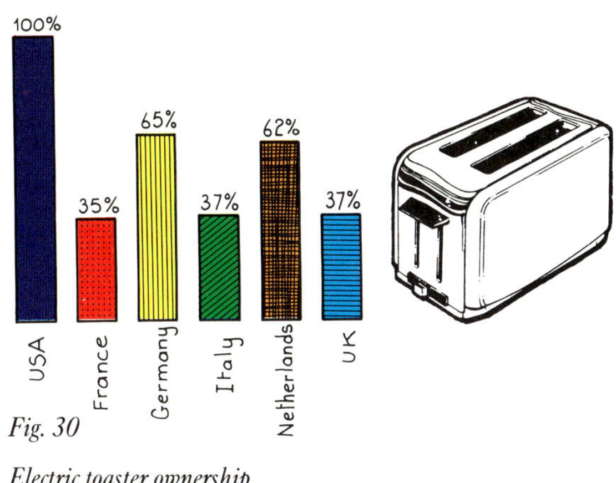

Fig. 30

Electric toaster ownership

4 How might the ownership levels shown in table 1 change between now and the year 2000?

Now study figs 27, 28, 29, 30 which give information about the ownership of electrical appliances in six different countries. Figures given relate to the percentage of domestic customers.

5 a Which country has the highest ownership of electrical appliances?
b What does this tell us about that country?

6 a Which three countries have very few electric cookers?
b Suggest alternative methods of cooking which these countries may use.

7 a List the ownership figures of electrical appliances for Italy.
b What do you notice about the ownership figures for Italy?
c Explain the high ownership of refrigerators in Italy.

Switch on again

1 Find out who is responsible for public electricity supply in England and Wales.

2 Most consumers pay their electricity bills four times a year or 'quarterly'. Write a letter to your electricity board asking for information on budget schemes to help you pay your bills.

3 a In 1880 Sir Joseph Swann's house at 99 Kells Lane, Gateshead, was the first house in Britain to be lit by electrical filament lamps. Find out about Sir Joseph Swann and write a short summary of his life.
b An early advertisement, which appeared in 1921, for electric lighting is shown. Imagine you work for The Electrical Development Association in the early twentieth century. Design an advertising poster to sell the idea of electricity to the public.

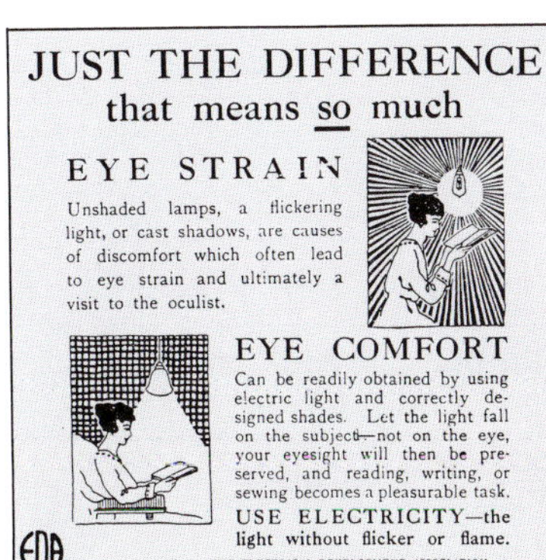

1921 advertisement for electric lighting

Cal Horrific

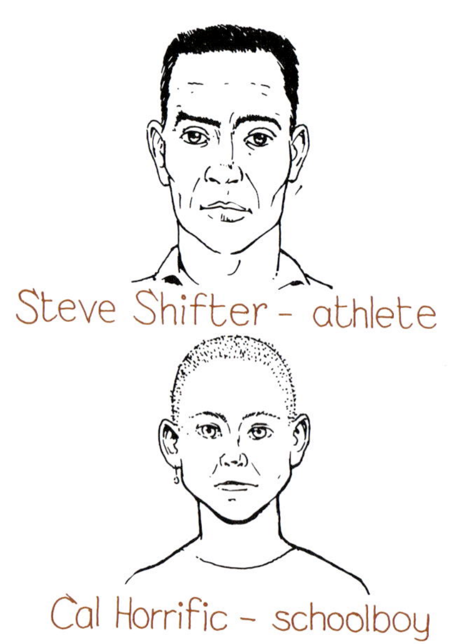

Steve Shifter - athlete

Mark Windbag - teacher

Susie Teabreak - bricklayer

Cal Horrific - schoolboy

Lisa Moaner - teacher

Di Etting - schoolgirl

Fig. 31

Cal's pals

name	intake
Steve Shifter	19,000
Mark Windbag	14,000
Susie Teabreak	13,000
Cal Horrific	13,000
Lisa Moaner	12,000
Di Etting	11,000

Fig. 32

Comparing energy intakes, kJ

1 Fig. 31 shows six inhabitants of Dualtown. Their daily energy intakes are given in fig. 32.

a Whose energy intake is the greatest?

b Whose intake is the least?

c Calculate the difference in intake between Mr Windbag and Mrs Moaner. Suggest a reason for the difference.

d Explain why Susie Teabreak's intake is greater than that of Lisa Moaner.

e Suggest reasons for Steve Shifter's large energy intake.

name	occupation	daily energy intake/kJ
Steve Shifter		
Mark Windbag		
Susie Teabreak		
Cal Horrific		
Lisa Moaner		
Di Etting		

Table 1 Daily energy intake of Cal's friends

2 Copy and complete table 1. Use the information given in both figs. 31 and 32 to answer this question.

3 a Write down Lisa Moaner's daily energy intake.

b Calculate her average hourly intake of energy.

4 Fig. 33 shows Cal Horrific's daily energy intake from the age of one year. Di Etting's energy intake over a similar period is given in table 2.

a Plot a graph to show how Di's daily energy intake changed over the years.

b From your graph estimate Di's energy intake when she was fifteen.

c At what age did Cal's energy intake begin to be larger than Di's?

age/years	daily energy intake/kJ
1	4 000
2	5 000
4	6 000
8	7 500
12	8 500
16	11 000

Table 2 Di's daily energy intake

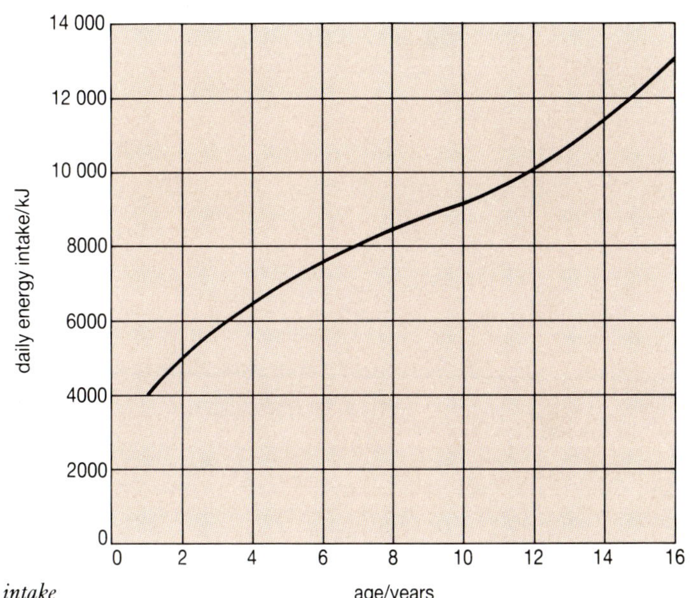

Fig. 33

Cal's daily energy intake

machine	efficiency
electric motor	80%
steam turbine	40%
fit human	27%
motor car	25%
unfit human	16%
steam engine	15%

Table 3 Comparing efficiencies in Dualtown

5 Table 3 compares the efficiencies of some machines in Dualtown.

a Calculate the difference in efficiency between a fit and an unfit human.

b Write down the efficiency of an *averagely* fit human.

c How does the efficiency of an average human compare with that of an electric motor?

d Draw a bar chart to display the information in table 3.

6 Answer this question in groups of four students. Discuss and list the factors which may decide how much energy you need. Draw a poster starring either Di Etting or Cal Horrific suggesting how they might use their energy.

Chew over these

1 Comment on the following statements.

- When sleeping a woman uses 4.2 kJ and a man 4.5 kJ of energy every minute.

- When sitting a man uses 5.8 kJ and a woman 5.0 kJ of energy every minute.

- When doing light housework a man uses 17.6 kJ and a woman 14.6 kJ of energy every minute.

Rocket sled

1 Copy and complete the passage on the jet engine. Choose your answers from the box.

> | | | | |
> |---|---|---|---|
> | reaction | forward | burnt | compressor |
> | eject | air | action | fuel |

Large amounts of . . . . , from the atmosphere, are drawn into a jet engine. A . . . . now compresses the air which passes to a combustion chamber where it mixes with the . . . . . The fuel is now . . . . . Exhaust jets, at the back of the engine, ... the combustion products. The . . . . of the exhaust gases being ejected backwards causes a . . . . which moves the jet . . . . .

2 In a jet engine the hot gas combustion products are passed through turbine blades. The turbine blades rotate and drive the compressor. Draw a flow diagram to show how a jet engine works.

3 Look at fig. 34. It shows a simple type of rocket engine.
   a Write down two similarities between a rocket and jet engine.
   b Write down two differences between a rocket and jet engine.
   c List two uses of rocket engines.
   d List two uses of jet engines.
   e Write a short passage to explain how a rocket engine works. You can assume there is a combustion chamber in the rocket engine.

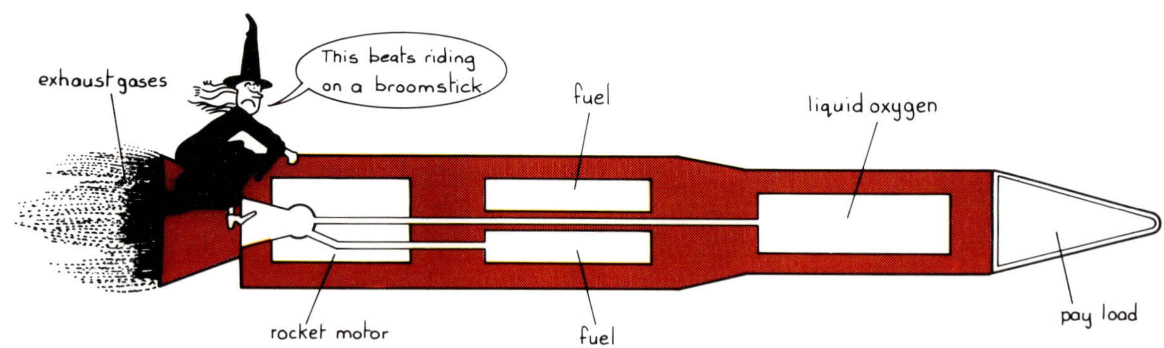

Fig. 34

Rocket engine, liquid hydrogen fuel

speed/$\frac{m}{s}$	0	168	280	280	280	0
time/s	0	3	5	10	15	16.5

Table 1 Rocket sled speed

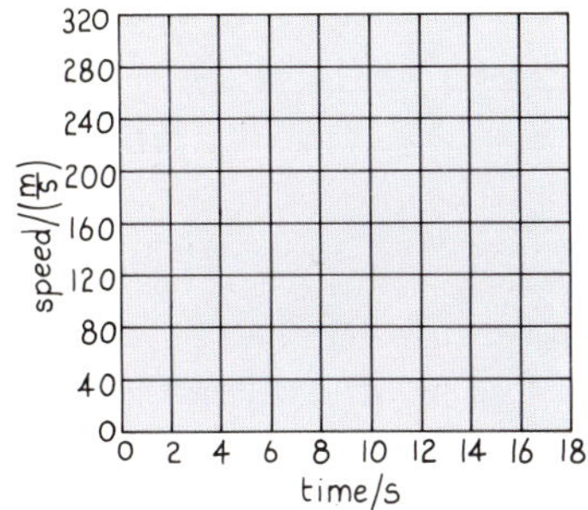

Fig. 35

Racing against time

4 Table 1 gives information about an experimental rocket sled which runs on rails.

a Plot a graph of speed against time. Copy the grid in fig. 35 on graph paper to start with.

b Briefly describe the motion of the rocket sled from your graph.

c How long did it take to reach its maximum speed?

d Calculate the:

(i) distance travelled in the first five seconds;

(ii) distance travelled at maximum speed;

(iii) distance travelled while slowing down;

(iv) total distance travelled by the rocket sled.

You can work out the distance travelled by calculating the area under the graph.

e Find the slope, or gradient, of your graph in the first five seconds. This is the acceleration of the sled in m/s^2.

f Find the force accelerating the sled in the first five seconds. The rocket sled's mass is 1000 kg. Your answer is in N.

$$force = mass \times acceleration$$

5 Answer this question in groups of four students. Imagine you are a team of engineers planning to break the world land speed record. You want to be sponsored by a large multinational jet engine corporation called Jetjag. You must persuade them to give you lots of money for your world record attempt, fig. 36.

a Discuss and decide upon a presentation to Jetjag.

b Draw a poster outlining your plan of action to Jetjag.

c Imagine your class to be the board of Jetjag. Give your presentation to the class. Try to convince them that they should sponsor your world record attempt!

High fliers

1 Find out about trams.

● What were their advantages?

● What were their disadvantages?

● Might they have a place in the twenty-first century?

Devise a tram system for your local area. Your system could be horse drawn, steam or electric.

2 The USA and USSR have spent huge amounts of money on space research. Much of this money has gone into the development of rockets. One point of view is that this money has been well spent.

a List the arguments for and against this point of view.

b Do you agree or disagree with the viewpoint?

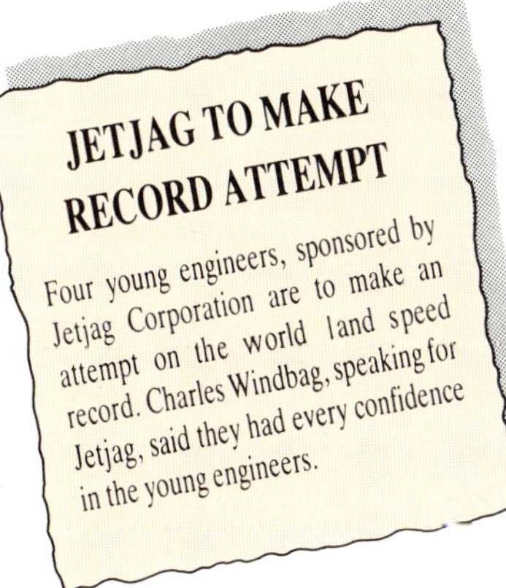

Fig. 36

Asking for money

Flying

Flight has always fascinated people. A reckless Benedictine monk, Oliver of Malmesbury, tried to fly in the eleventh century. He jumped from the tower of his abbey. Unfortunately, he was crippled as a result of his attempt. He was convinced that his failure was due to the lack of a tail in his design!

The Frenchman Alphonse Pénaud built a model aeroplane in 1870, fig. 37. Its propellor was driven by elastic and it flew without either nosediving or wobbling. The model flew a distance of 40 metres in eleven seconds on its first flight. It consisted of a thin piece of wood almost 1 metre long with wings about 0.85 metres wide. The wings were placed just in front of its middle. Its propellor was at the rear and was driven by twisted elastic. By bending the tip of the wings slightly upward, fig. 38, he gave it side-to-side stability. Pénaud gave his model front-to-back stability by bending its tailplane upwards, fig. 39.

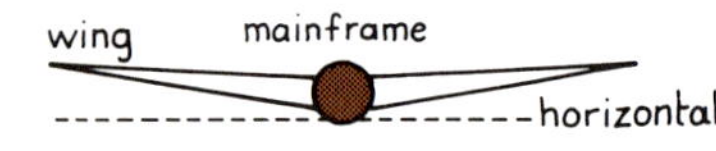

Fig. 38

Keeping it steady, side-to-side

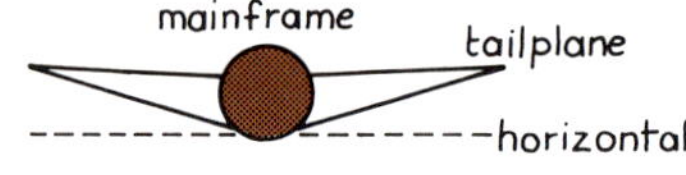

Fig. 39

Keeping it steady, front-to-back

1 a Why has man always wanted to be able to fly?

b Suggest reasons why Oliver of Malmesbury failed in his attempt to fly.

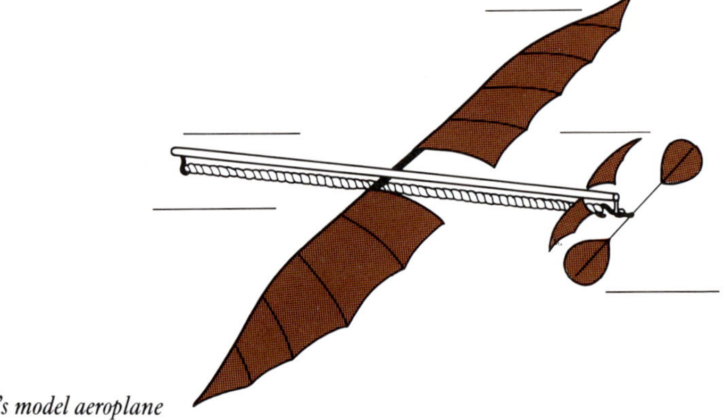

Fig. 37

Alphonse Pénaud's model aeroplane

2 a Copy and label fig. 37. Choose your labels from the box.

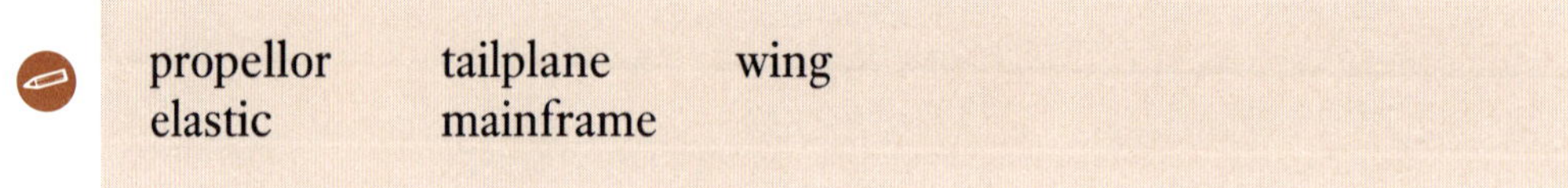

propellor	tailplane	wing
elastic	mainframe	

3 a How did Pénaud propel his model? Use a flow diagram in your answer.

b Explain how side-to-side stability was maintained in the model.

c How did Pénaud stop his model nosediving?

4 **a** What distance did Pénaud's model travel on its first flight?
b For how long was the model in the air?
c Calculate its average speed on its first flight.
d Why did the model aeroplane eventually stop?

5 Answer this question in groups of two students. Paper gliders are easy to make! Two simple gliders with different wing areas are compared in fig. 40. Will the glider with the large wing area stay in the air longer than that with the smaller wing area? Design an experiment to answer the question.
a What measurements will you take?
b List the stages in your experiment.
c How can you be certain you give each glider the same push when testing?
d What result do you expect? Why?

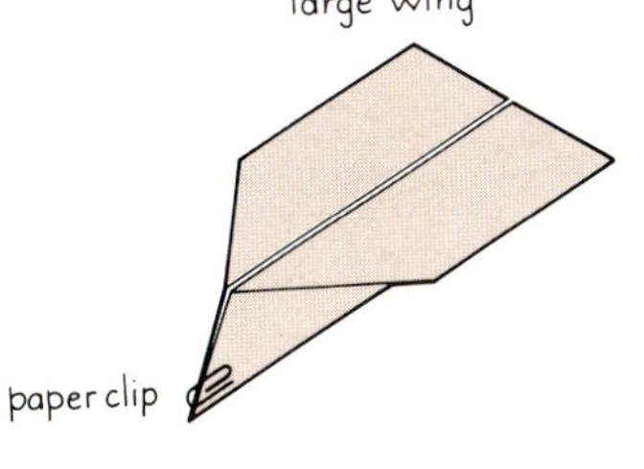

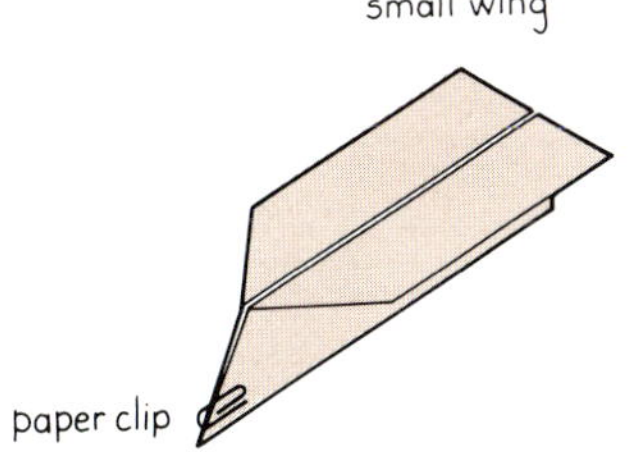

Fig. 40

Large and small wing areas

Low fliers

1 H.G. Wells was a great writer. In 1901 he suggested that people would fly before the year 1950. At the time his suggestion was treated with disbelief by the public! Two years later, on 17 December 1903, the Wright brothers made their first flight. Imagine yourself to be H.G. Wells in 1901. Write a letter to *The Daily Nag* newspaper arguing about the possible future benefits of flying.

Orville Wright 1911

2 Carry out your glider experimental design. Do gliders with large wing areas stay in the air longer than those with small wing areas?

3 Find out about the following people:

- Sir George Cayley, a Yorkshire squire who lived between 1773 and 1857

- Daedalus and Icarus

- Otto Lilienthal, a German who lived between 1848 and 1896

- Leonardo da Vinci, an Italian who lived between 1452 and 1519.
  What connections do they have with flying?

Double trouble

Copy the element grid, fig. 41. All answers to clues are names of elements. Solve the clues. Put the answers into your element grid. Read down the arrowed column. If your answers are correct you will see part of a newspaper headline of 1945.

Now copy fig. 43, the article grid. Transfer the numbered letters from your element grid to the article grid. The same number always means the same letter. Your article grid will give you the rest of the newspaper headline of 1945. Hopefully, you won't find this puzzle hard, as did George, fig. 42!

1. found in nuclear reactors
2. used to make microchips
3. inert, rhymes with jargon
4. rhymes with George's comment!
5. helps you breathe
6. metal, burns fiercely with white flame
7. halogen, group 7
8. . . . . ic acid
9. inert, ... lights
10. light, shiny metal, much used
11. precious metal
12. poison, group 5
13. another precious metal
14. same as clue 3
15. inert, was Superman born here?
16. halogen, used in antiseptics

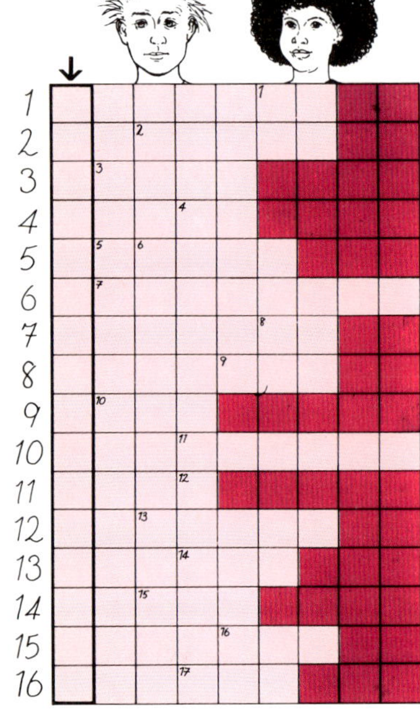

Fig. 41

Element grid

Fig. 42

George the thinker

Fig. 43

Article grid

Elementary, dear Watson

1 Find out about the newspaper headline of 1945.

- To what was it referring?

- Where did the event happen?

Write a short article to go with the newspaper headline.

2 a How many *names* of elements begin with each letter of the alphabet? Write your results in a table. Your table headings could be:

- letter

- number of elements.

Draw a bar chart to show your results.

b How many *symbols* of elements begin with each letter of the alphabet? Write your results in a table. Draw a bar chart to show your results.

3 a How many elements were named after famous scientists?

b Were any elements named after the scientists who discovered them?

4 Dmitri Ivanovitch Mendeléev was a Russian scientist who was born in 1834 and died in 1907. Find out and write about him.

Project solubility

The summer holiday was fast approaching and the students of class 4M at Midshire School were getting restless. Mrs Jones, their science teacher, was also looking forward to the holiday. She sensed her students' restlessness and decided to set them a project. The project was to take up the last week of term. Hopefully, this would stop 4M raising the roof in her lessons!

Mrs Jones organised 4M into two teams of scientists. Both teams were given the same task. Each team received powdered samples of four different salts. Their task was to investigate the solubililty of the salts at different temperatures. Their teacher explained that they could use any equipment in the laboratory. Mrs Jones asked them to write a brief report to include:

- a list of equipment used

- a brief description of what they did

- a set of results in the form of a table.

Fig. 44

Team A's report

1 Study team A's report, fig. 44.
    a What safety equipment did they use?
    b What is missing from their apparatus list which would allow them to use their Bunsen equipment?

c   What liquid did they use to dissolve their salts?

d   Were they correct to assume that each spoonful of salt was 1g? Explain your answer.

e   Was their method accurate?

f   Draw a labelled diagram to show how their apparatus might have been arranged.

g   Do you think team A's project was fair? Explain.

h   Draw a flow diagram to summarise how you think they carried out their investigation.

Team B   4M

Equipment bunsen, tripod, gauze, matches, different sized beakers, stirring rod, thermometer, measuring cylinder, distilled water, watch-glass, funnel, ice, electronic balance, pipette, graph paper, water bath, filter paper.

Method Measured out and put in 100cm³ of distilled water into a beaker. Put beaker into a constant temp. water bath at 20°C. Stirred in one of the salts until no more dissolved. Used pipette to draw out some solution which we put into a watch-glass which we weighed. Evaporated the liquid in the watch-glass to dryness. Weighed the mucky salt left in the watch-glass. Repeated the weighing and took an average.

Worked out how much salt in g was dissolved in original 100cm³ of water. Repeated, using the same salt, at different temperatures.

We repeated the investigation with the other salts and drew a graph of our results.

temp /°C	0	20	40	60
Sodium chloride/g	32	34	36	38
potassium chlorate/g	3	5	10	25
lead nitrate/g	38	53	70	90
potassium nitrate /g	10	30	65	110

Fig. 45

Team B's report

2   Look at team B's report, fig. 45.

a   Did they use any safety equipment?

b   Why did they use distilled water to dissolve their salts?

c   What is a saturated solution?

d   Draw a diagram to show how team B might have used the water bath.

e   What did they use the ice for?

f   How could they have evaporated the liquid in the watch-glass to dryness?

g   Was team B's project fair? Explain.

h   Was their project accurate?

i   Draw a flow diagram to show how you think they carried out their investigation.

Find the solution

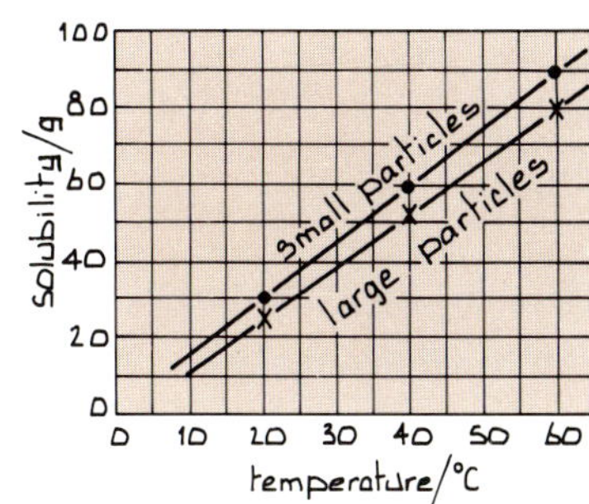

Fig. 46

Solubility and size of particle

1   Fig. 46 shows the results of a team of students. They investigated how the size of particle affected the solubility of an unknown salt Y.

a   What did they find out?

b   Can you explain their results?

c   How might they have measured the size of the particles?

Hair raising

1 Hair fashions change continually. Ancient Britons, two thousand years ago, loved to dye their hair in bright colours and grow it long. When the Normans invaded England over one thousand years later, the hairy Saxon residents must have been surprised to see the invaders sporting short hair!

    a List the hair styles of today which you think are fashionable.

    b Suggest reasons why hair fashions change continually.

    c All mammals have hair. Of what use is hair to mammals?

2 Two students, Sally and Jodie, carried out an investigation into hair growth. They measured the increase in length of their hair over eight weeks. Their results are shown in fig. 47.

    a Whose hair grew more quickly?

    b How long did it take Sally's hair to grow 18 mm?

    c What was the weekly increase in length of Jodie's hair?

    d Estimate Jodie's daily rate of hair growth.

3 Sally also measured the increase in length of one of her fingernails. Her results are shown in fig. 48.

    a How long did it take her nail to grow 6 mm?

    b What was the weekly increase in length of her nail?

    c Estimate the daily rate of growth of Sally's nail.

    d Describe briefly how she might have made her measurements.

4 a Sally counted the hairs in $1\,cm^2$ of Jodie's scalp. There were 173. She then estimated Jodie's scalp area to be $750\,cm^2$. Calculate the total number of hairs on Jodie's head.

    b How many hairs are on your head?

5 Look at fig. 49. It gives information about the elements present in hair.

    a Which element is the most common in your hair?

    b Draw a bar chart of the information given in fig. 49.

6 Design a simple experiment to find out how quickly your own hair grows.

    a What measurements will you make? What will you use to take measurements? How will you make your results accurate?

    b List briefly the stages in your experiment.

    c Suggest a table into which you might put your results.

7 a Hans Langseth grew his beard for 36 years. At his death his beard was found to be 533 cm long! The beard is still preserved in the United States. Estimate:

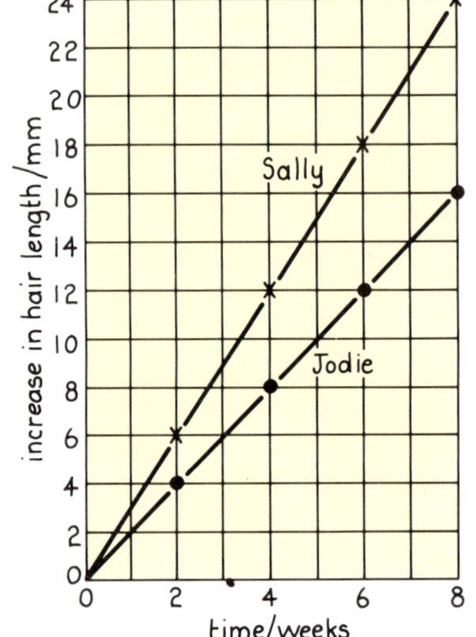

Fig. 47

Growing hair

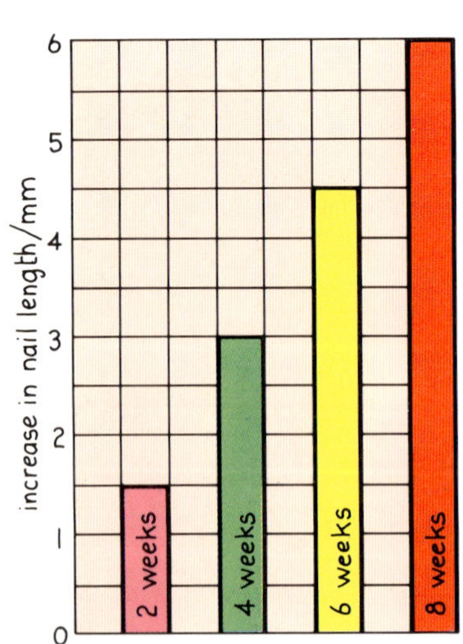

Fig. 48

Growing fingernail

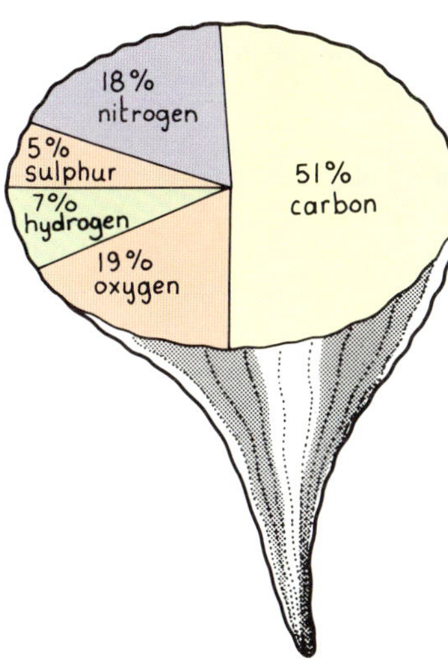

Fig. 49

Elements in hair

- the yearly growth, in mm, of his beard;

- his beard's daily growth in mm.

How does Jodie's hair growth compare to Langseth's beard growth?
b A Chinese priest took 27 years to grow a fingernail 58 cm in length! Estimate:

- the yearly growth, in mm, of his fingernail;

- his fingernail's daily growth in mm.

How does Sally's nail growth compare to that of the Chinese priest?

Splitting hairs

1 Look carefully at the photos. They show two human hairs.

  a Sketch both hairs.
  b Which is in good condition? Explain your answer.
  c Estimate the magnification of each photo.

Looking at hair

2 Discuss Jason's statement in the box.

My dad's totally bald. It's not going to happen to me. I'll look after my hair. I shall wash it every day and use hair restorer when I'm older.

Write a short passage agreeing or disagreeing with Jason.

3 Shampoos are generally alkaline. Cold rinses, such as weak lemon juice, are acidic. Hair is slightly acidic, with a pH between 4 and 6. Design a simple experiment to compare the effect of a weak acid and a weak alkali on hair.
  a What property or properties will you compare?
  b How will you test the pH of your acid and alkali?
  c Briefly describe what you will do.
  d Why should cold rinses be acidic? Do they do your hair any good?

Tired metals

metal	lifetime/years
iron	300
copper	73
lead	48
tin	39
zinc	37
gold	26

Table 1 Approximate world reserves of metals

1 Look at table 1. It gives information about known world reserves of some metals.

a Put the information in the form of a bar chart.

b Explain the meaning of the term *known world reserves*.

2 Many pure metals are soft and weak. They may be mixed together to form alloys.

a Table 2 is jumbled up. The *metals in alloy* and *use of alloy* are in the wrong rows. Copy and complete table 2 correctly.

b Which alloy appearing in table 2 consists of a metal mixed with a non-metal?

alloy	metals in alloy	use of alloy
brass	nickel, chromium	aircraft shells
bronze	aluminium, copper	heating coils
duralumin	iron, carbon	car bodies
mild steel	copper, tin	3–pin plug terminals
nichrome	copper, zinc	statues

Table 2 Uses of alloys

3 Professor Bender visited Alloy Academy to give a talk on metal fatigue. Jason Creep's notes on the Professor's talk are shown in fig. 50.

a Explain the term *fatigue failure*.

b List three everyday examples of metal fatigue.

c Do you think a fatigue failure starts inside a metal component or at its surface? Use a diagram to explain your answer.

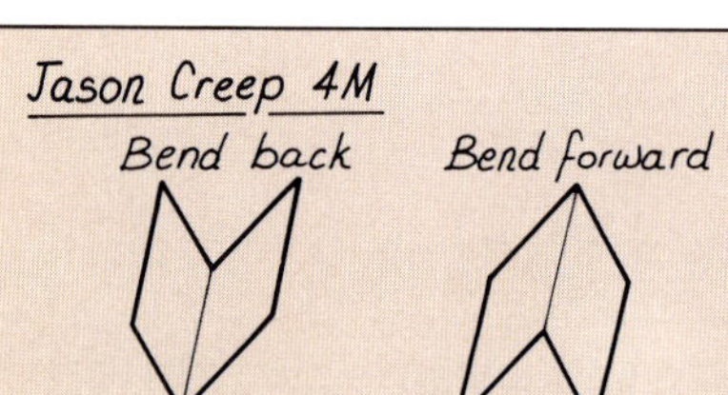

Fig. 50

Creep's notes

d   Professor Bender suggested large components fail before smaller scale components. What problems could this cause to aircraft and helicopters?

4   Answer this question in groups of two students. Imagine you have been given two packets of paper clips. One packet contains large and the other small paper clips. All the clips are made of the same material. You are asked to discuss and devise a simple experiment. Your experiment will allow you to find out the effect of size of paper clip on fatigue failure. You might wish to consider the following points.

- How will you bend the clips?

- What measurements will you take?

- How will you make the experiment fair?

- What equipment, if any, will you need?
  a   Draw a flow diagram to show the stages in your experiment.
  b   List the equipment you need.

Mind stretchers

hardness	mineral
1	talc
2	gypsum
3	calcite
4	fluorite
5	apapite
6	orthoclase
7	quartz
8	topaz
9	corundum
10	diamond

Table 3 Hardness scale

1   Table 3 shows a hardness scale. The scale is useful to mineralogists. On this scale materials can be classified according to their ability to scratch each other. Most metals are in the range of 3 to 8 on the scale.
    a   What is a mineral?
    b   Which is the hardest mineral on the scale?
    c   Where might your fingernail come on the scale?
    d   Imagine you have been given a lump of copper. Describe a simple test to find its hardness. Use a diagram to explain your answer.
    e   Where might copper come on the scale?

2   Surgeons can sometimes replace a human joint with an artificial joint. A human joint might be described as an excellent bearing.
    a   Why should human joints need replacing?
    b   List three joints that might be replaced.
    c   What materials might be used to replace a joint?
    d   What problems could arise from artificial joints?
    e   Draw a labelled diagram of one human joint.

3   Imagine you are a scientist working for a North Sea oil exploration company. Design a simple experiment to find out how metal fatigue affects the undersea parts of an oil rig. You might:

- use a paper clip to represent the rig

- use a solution of common salt to represent the sea.
  a   Draw a labelled diagram of the equipment you would use.
  b   Explain briefly how you would carry out your investigation.

Peruvian copper

Roman bronze

Copper is one of the most important metals in the world. It can generally be made into all kinds of shapes. Copper conducts both heat and electricity well. It doesn't rust easily and can be welded or soldered. Copper is often found pure in nature. For this reason it may have been the first ever metal used by man. Early copper objects broke easily. This was because the metal contained bits of stone and earth. Most copper exists in the form of ores which are found in rocks.

Over 5000 years ago the ancient Egyptians mined for copper ore in Sinai. From about 3500 BC men had been smelting the ore. Smelting involves heating the ore strongly to produce copper. It was common for tin and copper to be mined in the same area. People discovered that by mixing or alloying the two metals, bronze could be made. The ancient Greeks and Romans made huge amounts of bronze. Cyprus provided the Romans with their best quality copper. The word copper comes from the latin *cuprum*, meaning 'from Cyprus'. In Britain the Romans mined copper in North Wales and Cumberland. They alloyed the copper with tin from Cornwall to make bronze. Until the end of the nineteenth century nearly all the world's copper was smelted at Swansea in Wales. Britain's industry then declined. North America, Chile, Australia and Zambia are major producers of copper today. Most of the copper is now smelted near the mine. This is cheaper than transporting it overseas.

Ores may contain less than 4 per cent of copper. Nowadays the ore is crushed into small lumps. The lumps are ground to a powder in a ball mill. The copper is then separated out in a flotation machine. This increases the copper concentration to about 25 per cent. The ore is now smelted or heated strongly in a furnace. This produces copper of about 95 per cent purity. Finally, the copper may be further purified by electrolysis.

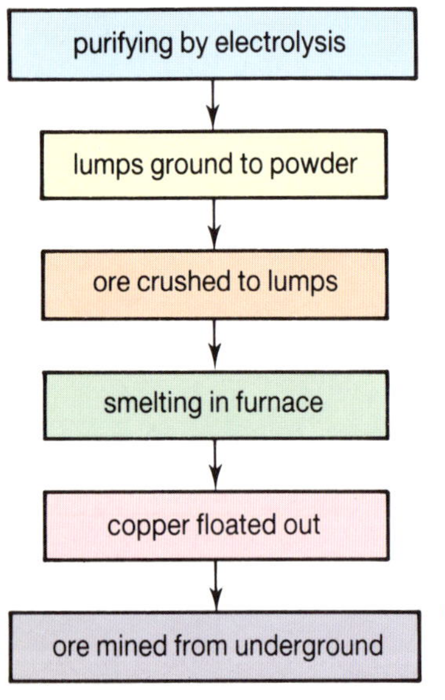

Fig. 51

Modern manufacture of copper

1 a Why may copper have been the first metal used by man?
b Why did ancient copper objects break easily?
c What is bronze?
d Explain where the word 'copper' comes from.
e What percentage of copper is contained in its ores?

2 a Why did Britain's large copper industry decline?
b List four modern, major producers of copper.
c List the properties of copper which make it an important metal.

3 Fig. 51 shows the jumbled up stages in the modern manufacture of copper. Re-draw fig. 51 showing the stages in the correct order.

4 Read Professor Copperhead's letter to Captain Silver, fig. 52. It concerns an ancient Peruvian industry. Peru is in South America.

a What is an archaeologist?

b Explain the term *prill*.

c When did copper smelting begin at Cerro de Los Cementerios?

d How long did the industry last?

e Explain why the ancient Peruvians chose the site they did.

f Draw a diagram to show the arrangement of furnaces.

g Use a flow diagram to describe how copper was produced.

h Why might the industry have died out?

i Why was the discovery of a copper industry great news to archaeologists?

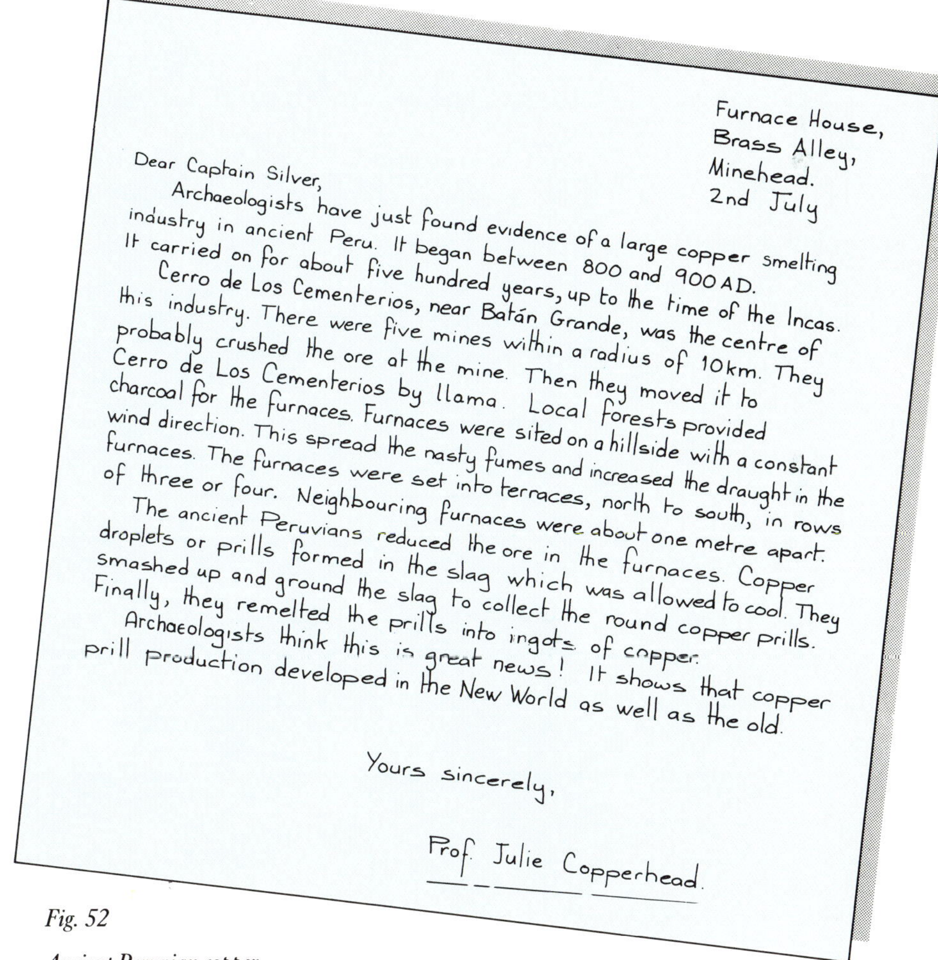

Furnace House,
Brass Alley,
Minehead.
2nd July

Dear Captain Silver,

Archaeologists have just found evidence of a large copper smelting industry in ancient Peru. It began between 800 and 900 AD. It carried on for about five hundred years, up to the time of the Incas. Cerro de Los Cementerios, near Batán Grande, was the centre of this industry. There were five mines within a radius of 10km. They probably crushed the ore at the mine. Then they moved it to Cerro de Los Cementerios by llama. Local forests provided charcoal for the furnaces. Furnaces were sited on a hillside with a constant wind direction. This spread the nasty fumes and increased the draught in the furnaces. The furnaces were set into terraces, north to south, in rows of three or four. Neighbouring furnaces were about one metre apart.

The ancient Peruvians reduced the ore in the furnaces. Copper droplets or prills formed in the slag which was allowed to cool. They smashed up and ground the slag to collect the round copper prills. Finally, they remelted the prills into ingots of copper. Archaeologists think this is great news! It shows that copper prill production developed in the New World as well as the old.

Yours sincerely,

Prof. Julie Copperhead.

Fig. 52

Ancient Peruvian copper

Getting down to brass tacks

1 Answer this question in groups of four students. Imagine you are a team of government advisers. Your government is worried that world copper supplies are running out. You have been asked to suggest ways to deal with the problem. Discuss the problem in your group. Draw a poster to summarise the group's advice to your government.

Video fission

1 Jackie's class watched a video on radioactivity. The class made notes during the programme, which they wrote up for homework. Jackie's homework is shown in fig. 53.

  a What particles are found inside the nucleus?
  b What particles are found outside the nucleus?
  c List three types of radiation. Describe what they are.
  d From which part of an unstable atom does radioactivity come?
  e Which radiation has the greatest penetrating power?

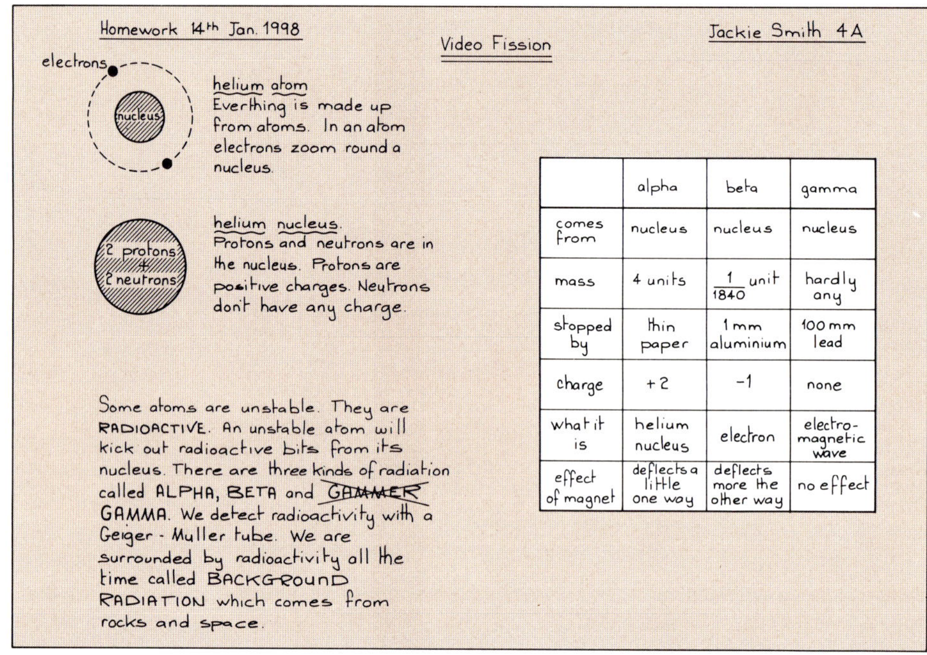

	alpha	beta	gamma
comes from	nucleus	nucleus	nucleus
mass	4 units	$\frac{1}{1840}$ unit	hardly any
stopped by	thin paper	1 mm aluminium	100 mm lead
charge	+ 2	−1	none
what it is	helium nucleus	electron	electro-magnetic wave
effect of magnet	deflects a little one way	deflects more the other way	no effect

Fig. 53

Jackie's homework

2 **a** Write down the mass of an alpha particle.
  b Write down the mass of a beta particle.
  c Calculate how much heavier an alpha particle is than a beta particle.
  d What is background radiation? Is it the same everywhere on Earth? Explain your answer.

3 Mrs Falzone, Jackie's teacher, demonstrated an experiment to the class. She wanted to find out if the activity of a radioactive source changed with time. The equipment she used is shown in fig. 54. Copy and label fig. 54. Choose your answers from the box.

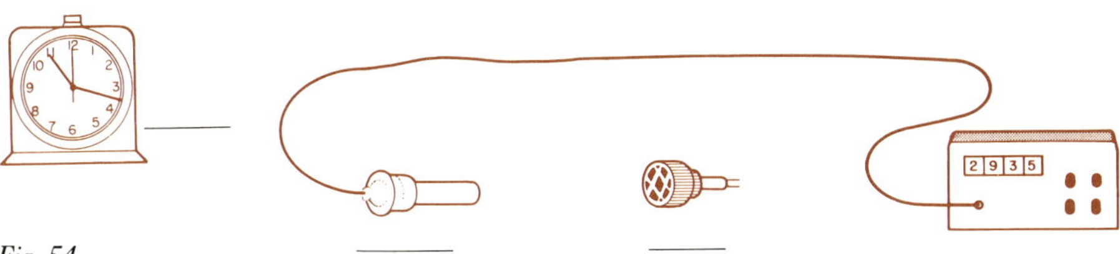

Fig. 54

Mrs. Falzone's experiment

source	detector	ratemeter	clock

count rate/counts s	time/min
1100	0
750	10
550	20
410	30
300	40
200	50
140	60
100	70

Table 1 Mrs Falzone's results

4 Mrs Falzone's results are given in table 1.
a Plot a graph of count rate against time. Copy the grid, fig. 55, to start with.
b At what time was the count rate 800 counts/s?
c At what time was the count rate 400 counts/s?
d How long did it take for the count rate to halve from 800 to 400 counts/s? Your answer tells you the *half-life* of the radioactive source.
e Briefly describe how you think Mrs Falzone carried out her experiment.

5 Answer this question in groups of four students. Imagine you are a team of nuclear scientists! You have been given a radioactive source. The source produces alpha, beta and gamma radiation. Discuss and design an experiment to separate the gamma radiation from the source. You may use any equipment you need.
a Draw a flow diagram to summarise the main stages in your experiment.
b List the equipment you require.
c How will you make your experiment fair?
d List the safety precautions you would take.

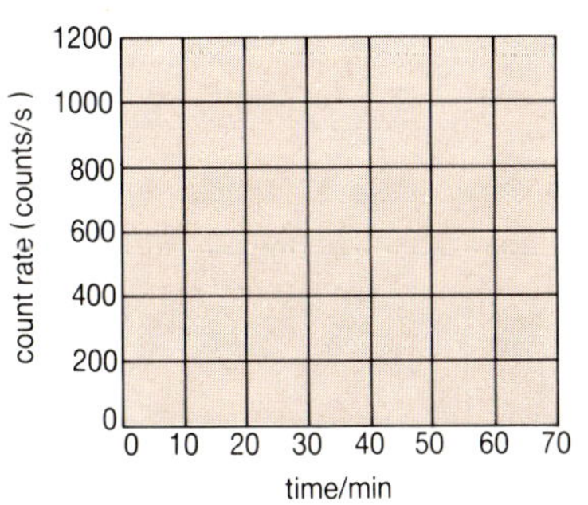

Fig. 55

Plotting your downfall

Curious queries

1 a Marie Curie, a brilliant scientist, was born in 1867 in Poland. She died in 1934 having made many exciting scientific discoveries. Find out and write about Marie Curie.
b The Becquerel family made many scientific discoveries in the nineteenth and early twentieth centuries. Antoine Cesar Becquerel lived between 1788 and 1878. His son Alexandre Edmond lived between 1820 and 1891. Alexandre's son Henri Becquerel has been called the father of radioactivity! Find out and write about the Becquerels.

Murder hunt

Experts comb horror house for new clues

FORENSIC experts will search a house today for clues to a triple killing.

Fingerprint checks will be made in every room of the house in East Dulwich, south London, where Mrs Cathy Dove, 35, and her children, Garey, 15, and Joanna, 13, were found shot dead on Friday.

Mrs Dove's husband, Charles, 53, was found barely alive and with a pistol near his side. He has had surgery and is too ill to talk to detectives.

Police have not ruled out the involvement of a professional killer.

Fig. 56

Triple killing

Ribs clue to corpse

THE murdered woman whose torso was found in a forest had a rib deformity which would have been known to a husband or lover, police said yesterday.

Sussex CID said her left ribs were more prominent. The torso of the woman, aged about 30, was discovered in Ashdown Forest on Sunday.

Fig. 57

Forest find

Fly swarm uncovers a murder

NEIGHBOURS who called in police to investigate a plague of flies sparked off a murder hunt yesterday.

A 29-year-old woman was found dead in the bedroom of her home in Brixton, south London.

Detectives believe Lorna Hayles put up a desperate struggle while being attacked by a large kitchen fork.

A detective said: "It was a very nasty murder."

But police have few clues at the moment.

They have been unable to find a motive — there was no sign of a break-in and none of the rooms were ransacked.

Last night teams of officers were carrying out house-to-house inquiries to try to discover when the murder was committed.

They were also trying to find out more about the woman who lived alone and had moved into the house last February.

Fig. 58

Murder hunt

1 In 1987 a triple killing occurred, fig. 56.

   a Where did the murders take place?

   b How were the victims killed?

   c What are forensic experts?

   d Why should they wish to search the house?

   e What else, other than fingerprints, might the forensic experts be looking for?

   f Look at your index finger. Sketch your fingerprint.

2 Study the newspaper article shown in fig. 57. It is about a horrific forest find.

   a What does the term *torso* mean?

   b Suggest why the police might have problems in identifying the victim.

   c What clue does the CID think may help identify her?

   d Imagine it had been you who had just found the body! Describe what you would have done next.

3 Read the article about the murder hunt, fig. 58.

   a Why were the police called to the scene of the crime?

   b Why might detectives have believed that the victim put up a struggle?

   c Explain the term *motive*.

   d Why did the police feel there was no obvious motive for the crime?

   e What were the police doing to find more clues to the murder?

4 Answer this question in groups of four students. Police often rely on the evidence of witnesses to catch criminals. How reliable are witnesses? Discuss and devise a simple experiment to find out. Use your classmates as witnesses. Don't forget to test their memories!

   a How will you test the memories of your witnesses?

   b Describe briefly how you would carry out your investigation.

   c Is it possible to make your experiment fair? Explain your answer.

   d Carry out your simple experiment. What did you find out?

   e How does a jury decide the reliability of a witness? List the factors which could influence them in their decision.

5 Write a short play about a murder investigation. The victim is male, middle-aged and has dark hair. Include the following clues in your play:

- bloodstained knife

- fair hair found on the victim's body

- half-eaten apple

- first-floor broken window near a drainpipe.

Act out your play in class.

Who dunnit

1 Study the newspaper article shown in fig. 59.

    a What mistake was made by the hospital?

    b Suggest how the mistake might have been made.

    c How did the hospital confirm their mistake?

    d Write a simple and brief set of guidelines for hospital staff. Your guidelines should ensure that a similar mistake is never made again.

2 Look carefully at the photo. It shows the scene of a common crime.

    a What crime has been committed?

    b List the clues the criminal may have left behind.

    c Imagine the crime has been committed in your classroom! Sketch the scene of the crime. It helps to label your sketch with the accurate position of each item of furniture.

Baby mix-up mother 'still in nightmare'

THE mother forced to hand back a baby after a hospital mix-up is still living a nightmare, her husband said last night.

Maureen McHugh took home the wrong baby after name tags were accidentally switched at Portlaoise hospital, near Dublin, Eire.

"Maureen is still very upset," said farmer Michael McHugh.

"She has now accepted that there was a mistake. But of course she is finding it difficult to live with giving up the son she regarded as her own."

The hospital mix-up was confirmed after both babies were given blood tests.

Mrs McHugh is expected to rejoin her 37-year-old husband and their other son on the family's farm at County Offaly, Ireland, within the next few days.

Mr McHugh said his wife's agony had been made worse because she had suffered two miscarriages.

He added that the other mother, Assumpta Broomfield, feared something was wrong on the day the children were born, August 21.

Mr McHugh also hinted at possible legal action.

Fig. 59

Hospital mistake

Criminal intent

Chips

Fig. 60

Pop

1
  a   List the names of three fizzy drinks.
  b   Name the gas, X, contained in the fizzy drink bubbles.
  c   Look at fig. 60. What happens to the limewater?
  d   Will a lighted splint burn in gas X?
  e   Is gas X lighter or heavier than air? Explain your answer.

2 Gas X is also produced when an excess of dilute hydrochloric acid is added to limestone, calcium carbonate. A team of scientists from Chiplump Chemicals investigated the reaction. They wanted to find out if the rate of gas produced was affected by the size of limestone used. Chiplump's scientists used three sizes of limestone. They did three experiments using *a large limestone lump*, *small chips* and *powder*, each of mass 2 g. They noted the volume of gas X produced every two minutes. Their results are shown in fig. 61.

  a   Which size of limestone reacted the fastest?
  b   Describe the shape of Chiplump's graphs.
  c   Copy and complete table 1 using the information given in fig. 61.
  d   What did the scientists find out?

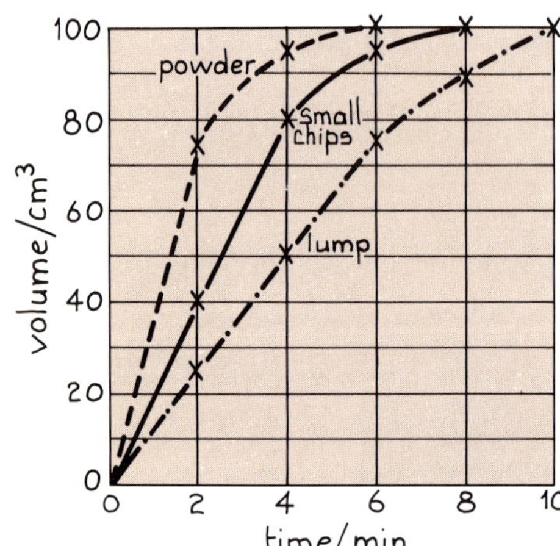

Fig. 61

Chiplump's results

time/min	powder volume/cm^3	small volume/cm^3	lump volume/cm^3
0		0	0
2			
4			50
6			
8	100		
10	100	100	100

Table 1 Chiplump's missing values

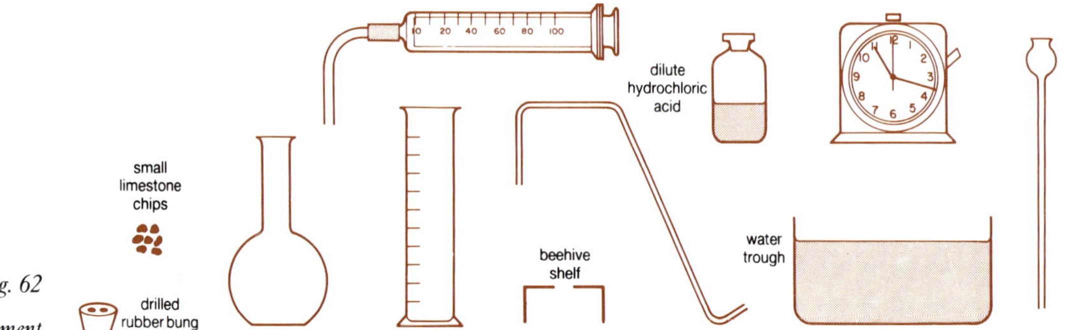

Fig. 62

Pick your equipment

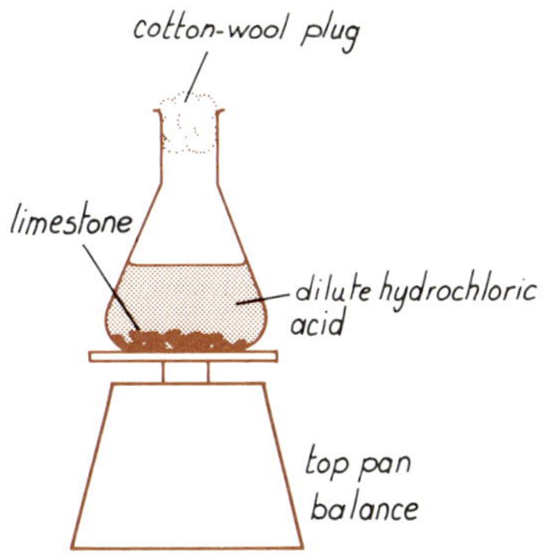

Fig. 63

Limebone's equipment

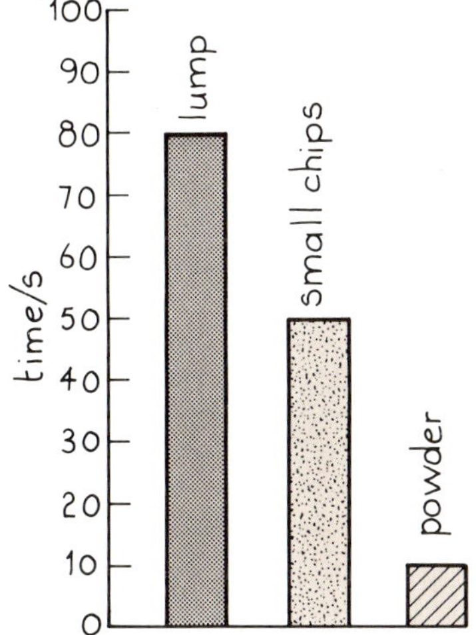

Fig. 64

Limebone's results

3 **a** Name the unlabelled items of equipment in fig. 62.

b The Chiplump scientists used some of the items shown in fig. 62. Which items might they have used in their investigation? Suggest other items of equipment, not shown in fig. 62, that the scientists may also have used.

c Draw a labelled diagram suggesting how Chiplump's equipment may have been set up.

d What safety precautions might the scientists have taken in their investigation?

4 Limebone Chemicals carried out a similar investigation. Limebone's scientists used the equipment shown in fig. 63. They used *a large limestone lump, small chips* and *powder*, each of mass 10 g. They noted how long the total mass on the balance took to decrease, each time, by 0.5 g. Their results are shown in fig. 64.

a Which of Limebone's experiments took the longest to complete?

b Why did they use a cotton-wool plug in their investigation?

c Suggest a reason for the decrease in total mass.

d Draw a flow diagram to summarise the stages in their investigation.

e What did Limebone's scientists find out?

5 Answer this question in groups of two students. Imagine you are scientists working for Chiplump Chemicals. You are investigating the reaction of hydrochloric acid with limestone. Discuss and design an experiment to find out if the strength of acid affects the rate of production of gas.

a List the equipment you will use.

b Describe briefly how you intend to carry out your investigation.

c How will you make sure your results are accurate?

When the chips are down

1 A student investigated the reaction of magnesium ribbon with an acid. A gas was given off which she thought may have been hydrogen. She collected the gas using a syringe and noted the volume present every two minutes. Her results are shown in table 2.

a What test could she have applied to identify the gas?

b Plot a graph of her results.

c Suggest reasons for the shape of the graph.

2 Limestone is one of the most common rocks found in nature. Its uses include making roads and building. Chalk is one soft form of limestone. Write brief answers to the following questions.

a Where is limestone found in Britain?

b How might limestone be formed from animal remains?

c What happens to limestone when it is burnt?

d How is cement made?

time/min	volume/cm^3
0	0
2	30
4	43
6	49
8	50
10	50

Table 2 Magnesium and acid

Slick experiment

TEACHER: Please keep quiet, fourth years. Today you're going to investigate. Shut up or you'll all get lines. Thank you. Get your finger out of that electrical socket Leigh, you could damage it. We'll be looking at three different oils. They are fuel oil, bike oil and multigrade engine oil. The fuel oil is from the school's central heating system. You are going to compare how thick or viscous the oils are. You might do this by finding out how quickly the oil fills a container. Are there any questions?

LEIGH: Yes Miss, do I have to work with Jane again? She's always picking on me.

TEACHER: I'm afraid Jane will have to put up with you again Leigh. Right, any more questions? Please collect your equipment and get started. Oh yes! Hang on a minute. Don't make a mess or you'll stay behind at break to tidy up.

1 a Look at fig. 65. It shows the equipment that Jane and Leigh used. Copy and label their apparatus.

b For what purpose might they have used their marker pencil?

c How did they prevent the oil from running directly into the measuring cylinder?

Fig. 65

Leigh's diagram

2 Their results are shown in fig. 66.

a What measurements did they make?

b Describe briefly how they might have carried out their experiment.

c Suggest why there are only two results for fuel oil.

d Jane's graph is incomplete. It shows only the results for engine oil. Copy and complete her graph to compare the results for all three oils.

e Which oil took the longest to fill its measuring cylinder? Suggest which of the oils is the most viscous.

| Jane Johnson 4u | Investigation into the thickness of oils | | |

volume in cylinder /cm³	time/s		
	fuel oil	bike oil	engine oil
0	0	0	0
2	–	3	6
4	–	6	12
6	–	9	18
8	–	12	24
10	1	15	30

Fig. 66

Jane's results

3 Answer this question in groups of two students. Engine oil gets hot in cars. Does the thickness or viscosity of the oil change as its temperature rises? Discuss and design an experiment to find out. You might choose similar equipment to that used by Jane and Leigh.

a List the items you might use. Sketch your experiment in action.

b What safety precautions should you take?

c Describe briefly what you would do. Design a table for your results.

More toil

1 A new car is being developed. Fig. 67 shows the materials used in the vehicle.

a Of which material is the car mostly made?

b What percentage of the vehicle's mass is unaccounted for?

c Suggest two materials, not shown in fig. 67, which might be present in the car.

d The car's total mass is 1000 kg. What mass of copper is in the vehicle?

e Design a colourful advertising poster for the car.

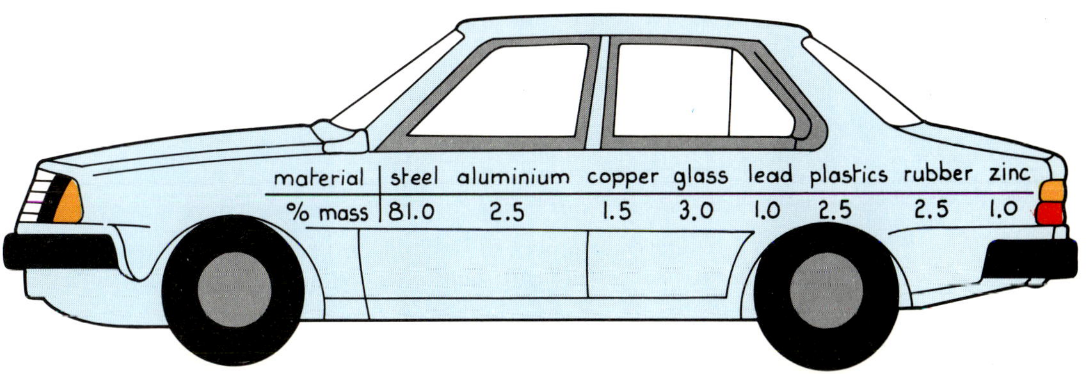

Fig. 67

Materials used

Ally Minyum

Bauxite is the main ore of aluminium. It is named after Les Baux, in France, where it was first found. Jamaica, India and West Africa are other countries with deposits of bauxite. Wohler, in 1826, was the first scientist to separate aluminium from its ore. He did this by heating or smelting the ore to over 2000 °C. This method was very expensive due to the high cost of heating and the small amounts of aluminium made. Early in the nineteenth century aluminium was more expensive than gold!

Héroult and Hall, in 1886, discovered another way of making aluminium. They passed electricity through aluminium oxide. The process is called **electrolysis**. Unfortunately, the electricity used to produce one tonne of aluminium could power your home for ten years. As a result most modern aluminium factories are sited near hydro-electric power stations to cut costs.

1   a   Write down the name of the main ore of aluminium.
   b   From where does the ore get its name?
   c   List three countries with mineral deposits of aluminium ore.

2   a   Use a diagram to explain how Wohler first produced aluminium.
   b   Why, in the early nineteenth century, was aluminium production expensive?

3   a   Name the modern process used to separate aluminium from its ore.
   b   Where should aluminium plants be sited?

4   Ally Minyum carried out an investigation into the electrolysis of copper(II) sulphate solution. Ally's notes are shown in fig. 68.
   a   List the equipment he used. It's not all in his diagram!
   b   How did he change the current through the copper(II) suphate?
   c   What did he use for the anode and cathode?
   d   Finish plotting Ally's graph. Use graph paper.
   e   Ally explained what happened by writing about copper and sulphate ions. What happened to the cathode in his experiment? Did anything happen to the anode?

5   Discuss this question in groups of two students. Imagine you can use the same equipment as did Ally. You may also use any other apparatus you might need. Discuss and design an experiment to find out what happens to the masses of the anode and cathode during electrolysis.
   a   Use a flow diagram to explain the stages in your design.
   b   Draw a labelled diagram to show your experiment in action.
   c   What safety precautions should you take?

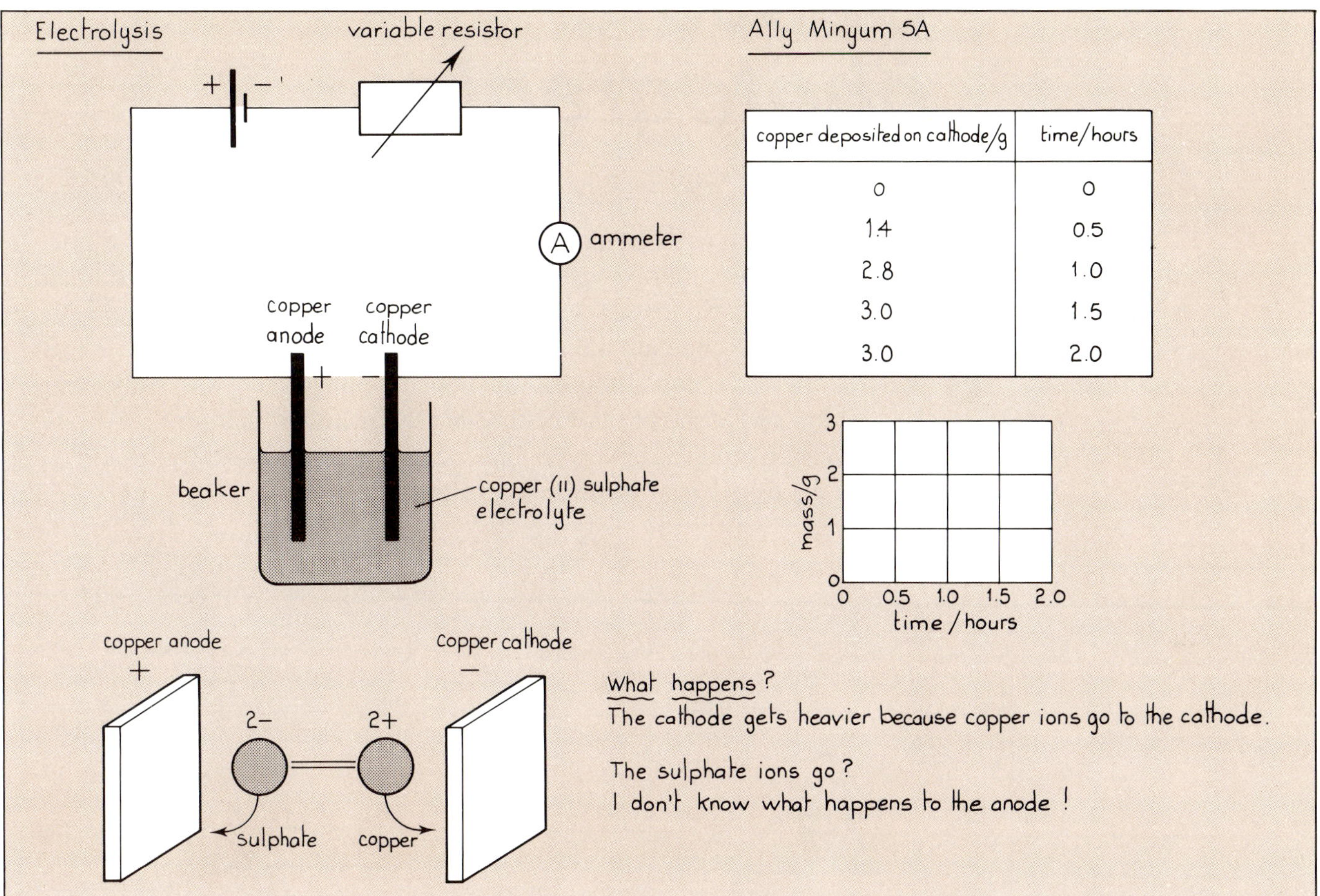

copper deposited on cathode/g	time/hours
0	0
1.4	0.5
2.8	1.0
3.0	1.5
3.0	2.0

Fig. 68

Ally's notes

Charge on

1 In 1886 two scientists, working separately, discovered the electrolytic method of producing aluminium. Charles Martin Hall lived in America. It is said the Frenchman Paul Héroult discovered the process two months after Hall. Find out and write about these two scientists.

2 a Write a short paragraph explaining how Ally Minyum might have made his experiment fair.
b Ally's explanation of what happens was not complete. Re-draw Ally's diagram, explaining what happens to the anode.
c How did Ally make his experiment accurate?
d How, do you think, is very pure copper made? What is very pure copper used for?

3 Carry out the experiment you designed. What did you find out?

Obscura photography

1 Fig. 69 shows an early type of camera obscura. It was a light-tight room acting as a huge camera. Light, from an object, travelled through a small hole in one wall. Sixteenth century observers, inside the room, could see an image on the opposite wall.

  a Copy and complete fig. 69 to show the image seen by the observers.

  b How might the observers have made the image sharper?

  c What would the observers have seen if a second hole was made close to the first? Draw a sketch to explain your answer.

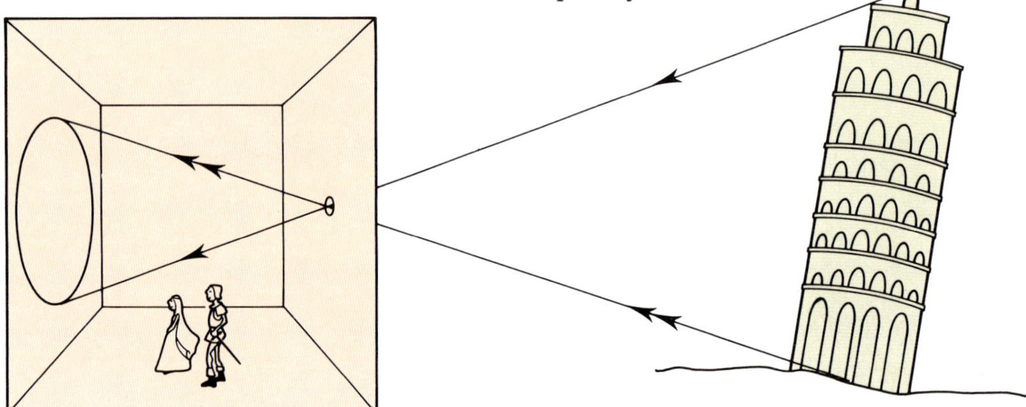

Fig. 69

Towering obscura-ty

  d Photochromic spectacle lenses change colour in bright light. List three other objects which change colour in light.

2 Smaller types of camera obscura were used by artists two hundred years ago. The camera obscuras helped the artists to trace landscape images, fig. 70. The cameras contained:

- a convex lens to focus the distant image

- a mirror to reflect the image onto the top surface.

  a Copy and complete fig. 70. Your final diagram should show how the image is tranferred to the top surface.

  b What materials did the artist need to build a camera obscura?

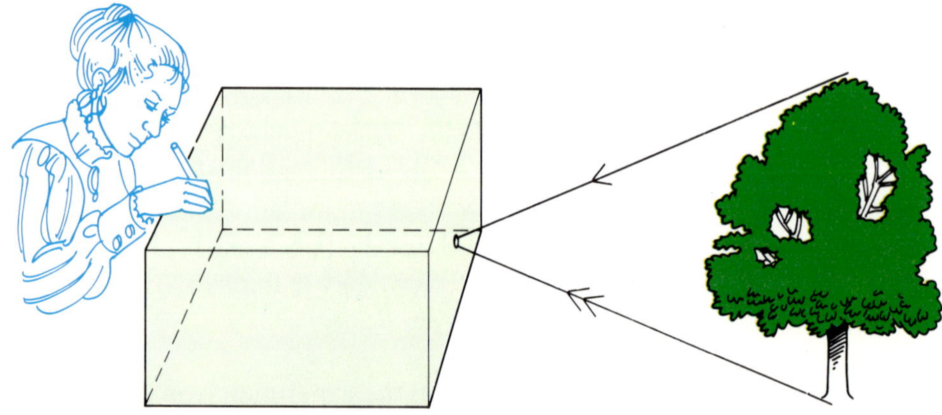

Fig. 70

Artistic licence

Louis Daguerre 1787 – 1851

William Fox-Talbot 1800–1877

Calotype photograph of Constance Talbot 1840

3 Joseph Nicéphore Niépce, born in France in 1795, is known as the inventor of photography. Using a camera he fixed a positive image of a subject onto a pewter plate. The process took several hours. The pewter, an alloy of tin and lead, had been covered in bitumen. Three years later, in 1829, Niépce formed a partnership with Louis Daguerre. Daguerre was a theatre scenery painter. They wanted to commercialise photography. Unfortunately Niépce died in 1833. Within six years Daguerre had developed the **daguerrotype process**. He coated a copper plate with light-sensitive silver nitrate. A single exposure followed by development produced a positive image. He then fixed or made permanent the image on the plate. This was done by soaking the plate in brine or hypo.

a When did Niépce first fix a positive image onto a plate?

b What is pewter?

c Why did Niépce and Daguerre form a partnership?

d Explain the photographical term *fix*.

e Draw a flow diagram to describe the daguerrotype process.

f Daguerrotypes had a huge impact on the world. Suggest changes that have taken place as a result of commercial photography.

g What jobs have been created following the invention of photography?

4 Answer this question in groups of four students. The Englishman William Fox-Talbot patented the **calotype process** in 1841. This was the first commercial positive/negative photographic system. He first made a flimsy negative paper print. His final positive paper print was called a calotype. By 1841 daguerrotypes had become popular. For this reason calotypes did not catch on quickly. However, Fox-Talbot's invention led eventually to modern photography.

Imagine you are living in 1841! You work for William Fox-Talbot. He wants you to sell his invention to the public.

a To what uses might the calotype process be put?

b At whom might you aim your advertising?

c Design an advertising poster to sell the invention. Where would you place the posters?

d What other forms of advertising might you use? Don't forget that it's 1841!

More obscura questions

1 Find out and write down the meanings of the words in the box.

agitation	aperture	developer	emulsion
film	highlight	iris	shutter

2 Write a short article about the modern uses of photography.

Soap opera

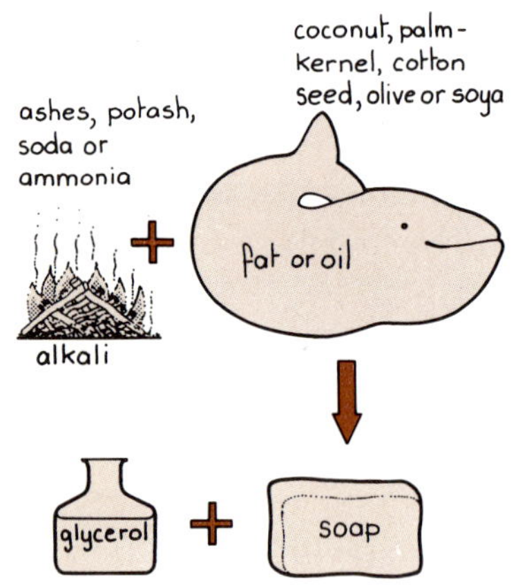

Fig. 71

Early soap making

'*The stain of your sin is still there and I see it, though you wash with soda and do not stint the soap*'. *Jeremiah, chapter 2, verse 22.*

The above quotation from the Bible suggests that man has known how to make soap from ancient times. Scholars believe that this quotation was written about 630 BC. Early man probably discovered soap by accident when ashes from his camp fire jumped into his frying pan containing fat! The ashes and animal fat combined to produce soap and glycerol, as shown in fig. 71.

The Romans were known to make soap by boiling goat's fat with the ashes of beechwood to produce soft potash soap. The Romans learned how to make soap from the Gauls, who had learned from the Phoenicians. This early soft soap was probably yellow in colour and very unpleasant to smell!

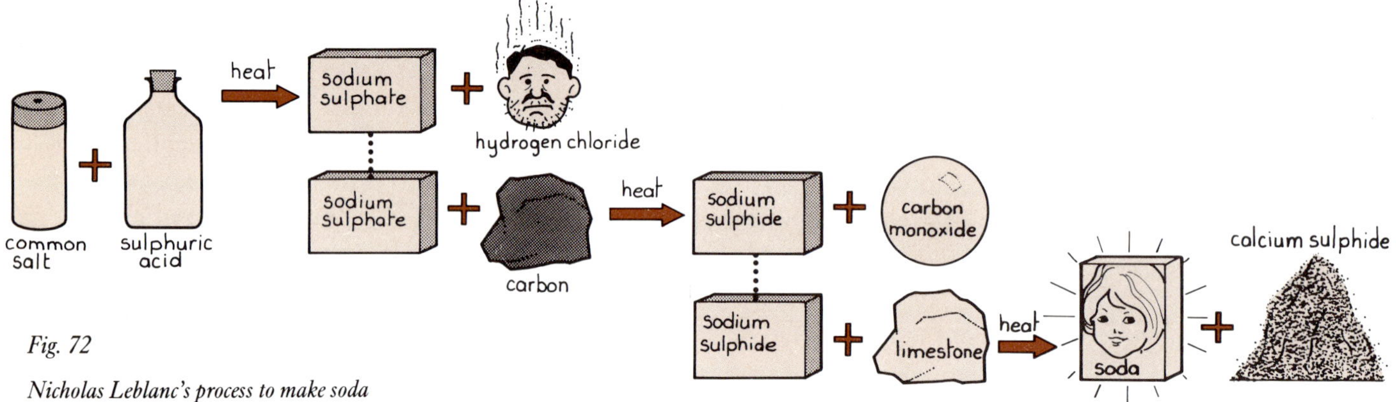

Fig. 72

Nicholas Leblanc's process to make soda

In the middle ages scented soaps were produced in France, Spain and Italy which were exported. However, only the very rich could afford these improved soaps. The vast majority of the population had to make do with the smelly, soft potash soap as made by the Romans.

Soap had been rare and expensive due to a lack of a good supply of alkali. However, in 1791 Nicholas Leblanc discovered a way of making soda from salt. This gave the soap-making industry a great impetus. His process is shown in fig. 72. Leblanc's process to make soda caused much hydrogen chloride gas pollution. Also huge piles of calcium sulphide waste were formed. To minimise the problem in the United Kingdom, Parliament passed the Alkali Act of 1863. This forced industry to absorb at least 95 per cent of the hydrogen chloride gas produced.

The Solvay, or ammonia-soda, process is a more modern method of producing soda. The advantages of this process over the Leblanc process are that it is almost self-contained and cheaper.

At the beginning of the twentieth century a severe shortage of animal fats occurred. Scientists managed to substitute vegetable oils for animal fats in the soap-making process. As a result of the plentiful supply of soda and vegetable oils, soaps become cheaper. More people were now able to buy soap. It became an everyday item rather than the expensive luxury it had been for thousands of years. Soap could now make an important contribution to the general hygiene of the population.

1 Copy and complete the passage below.

The . . . . learned how to make yellow, smelly soap from the Phoenicians. Later, in Europe, expensive, scented soaps were available only to the very . . . . . The discovery of the . . . . and Solvay processes helped to make the production of soap . . . . . In the . . . . century animal fats were replaced by . . . . oils in the soap-making process. This allowed soap to become an . . . . item.

2 a Write down three examples of alkalis.

b Write down three examples of oils.

3 What evidence is there for believing that man could make soap more than 2500 years ago?

4 How did the Romans produce soap?

5 Why, before the twentieth century, was soap rare and expensive?

6 a What discovery gave the soap-making industry a great impetus?

b What effect on the environment occurred as a result of this discovery?

c How was this environmental problem minimised in the United Kingdom?

Brainwashers

1 Estimate how much soap, in kg, is used in the United Kingdom every year. You may assume a UK population of 50 million people and that the average large bar of soap is 150 g. Assume also that the average person uses one such bar of soap every fortnight.

2 Discuss this question in groups of four students. Imagine you live in ancient Britain in the year 500 BC and have no soap! How would you keep yourself clean? Draw a poster to summarise your ideas.

3 Find out about the Alkali Act of 1863. Write brief notes on this Act of Parliament.

Do fabrics stretch?

Sharmina and John were asked to carry out an investigation during their science lesson. They were expected to write a report which was to be assessed by their teacher.

Sharmina and John collected a tray which contained their equipment, together with an instruction card. The instructions on the card are shown in fig. 73.

Sharmina thought it would be wise to look up the word *elasticity* in her textbook. 'Elasticity is the property of a body to go back to its original size

Fig. 73

Sharmina and John's instruction card

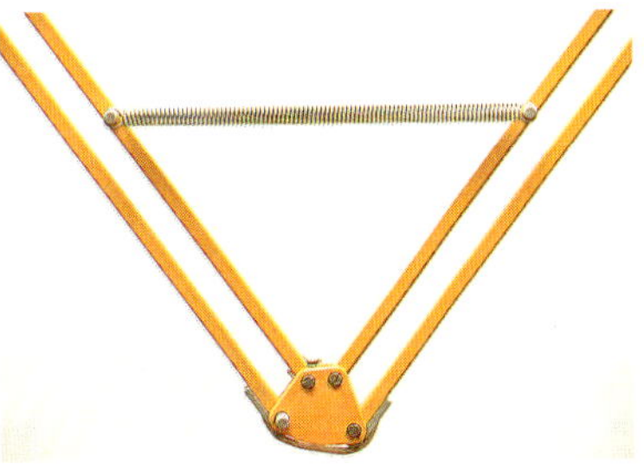

Testing a spring

and shape after being deformed', quoted the textbook. 'When a spring is stretched the molecules are pulled apart. When the stretching force is taken away the force of attraction is so strong between molecules that they go back to their original positions. The spring is said to be elastic.'

John was sure an elastic band must be elastic. They agreed to stretch an elastic band to see if it was. Sharmina measured the elastic band's original length, then stretched it with her hands. She then released the elastic band slowly. To their amazement they found that its final length was not the same as its original length!

1 a What did Sharmina use to measure the elastic band's length?

b Was Sharmina's method of applying stretch sensible? Explain your answer.

c How would you have applied stretch to the elastic band? Use a labelled diagram in your answer.

d Did the elastic band increase or decrease in length after stretching?

e Was the elastic band behaving elastically?

2 Sharmina and John listed the following questions which, they thought, could help them in their investigation.

a Shall we test single fibres from the fabric or pieces of fabric?

b What shall we use to measure the length of fabric before and after stretching?

c What can we use to measure the diameter of a single fabric fibre?

d What other equipment will we need?

e What will our testing equipment look like when it is set up?

Can you answer their questions?

Stretch your mind

1 Discuss this question in groups of four students. You are the owners of a small sportswear business, designing and manufacturing sports clothing! You wish to choose the best material for making tennis tops. Before you make your final choice of material you must decide which *factors* are important. Some factors you may wish to think about might be:

the consumer	the price	the weight
the colour	stretchiness	wear

Work out and write a questionnaire to allow you to choose the best material for making tennis tops.

Boris Becker – stretching for a serve

2 Carry out Sharmina and John's investigation. What did you find out?

Waste matters

1   Getting rid of rubbish can be a problem. Some is used to help reclaim derelict land. Other waste may be burnt in incineration plants. These plants are open 24 hours per day. Fig. 74 shows how solid waste is incinerated in such a plant.
   a   Explain the term *incineration*.
   b   Describe the energy changes that take place in the plant.
   c   Write a short summary to describe how the plant operates.
   d   List two advantages of burning rather than burying waste.
   e   Rubbish is burnt at a temperature of about 1000 °C. What problems might this high temperature cause?

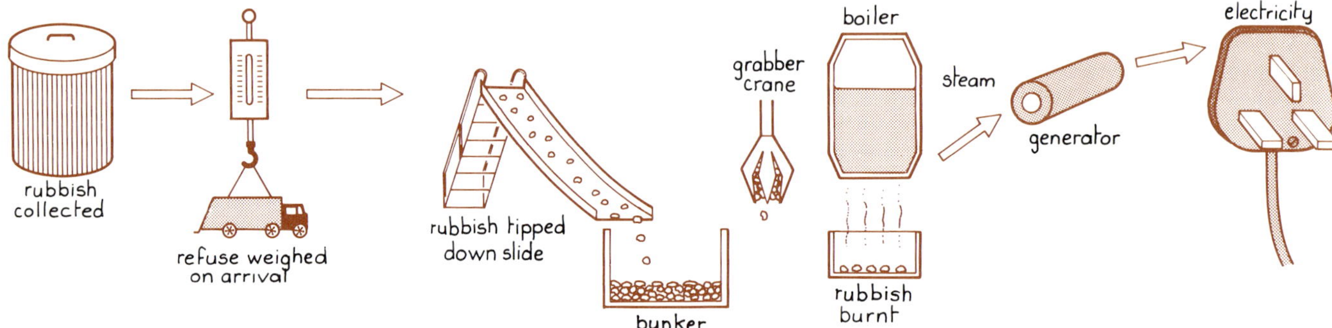

Fig. 74

Burn that rubbish

2   The content of refuse varies quite a lot. Fig. 75 gives details of refuse arriving at an incineration plant.
   a   What forms the largest part of the refuse?
   b   How much of the refuse arriving is *unclassified*? Suggest two examples of material that might come into this category.
   c   Copy and complete table 1. Use the information given in fig. 75.
   d   Draw a bar chart of the details in your table.

material	% weight

Table 1 What a load of rubbish!

Fig. 75

Raw refuse, percentage weight

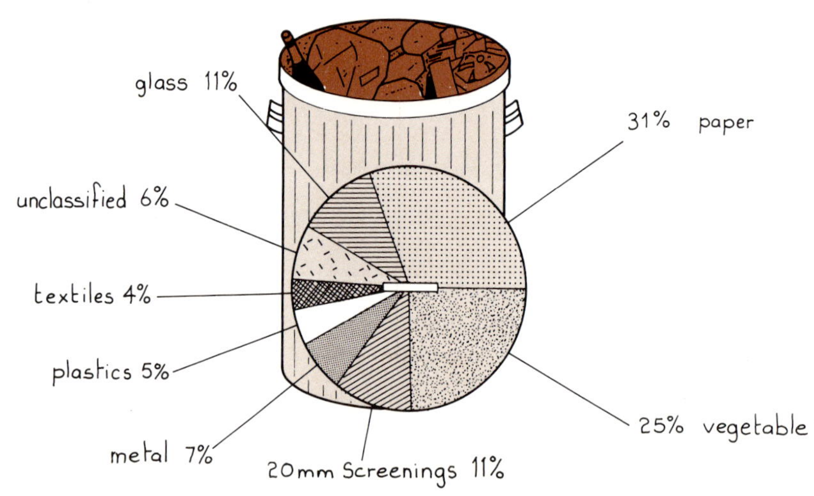

element	% weight
cadmium	0.01
calcium	9.15
chromium	0.03
copper	1.27
iron	15.00
lead	0.83
magnesium	2.25
manganese	0.54
nickel	0.06
sodium	2.60
zinc	1.18

Table 2 Residual composition, metals

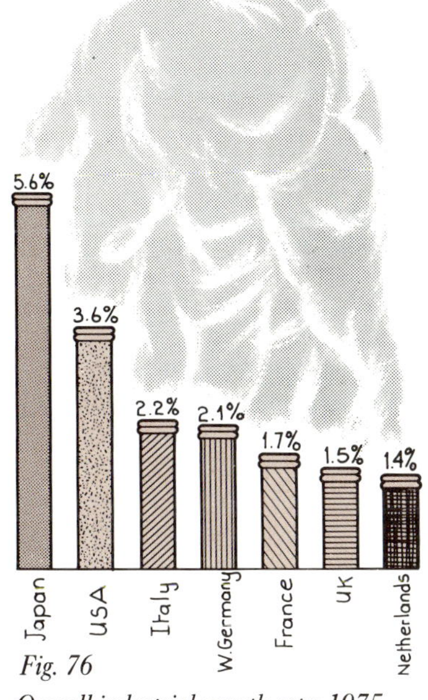

Fig. 76
Overall industrial growth rates 1975–1985

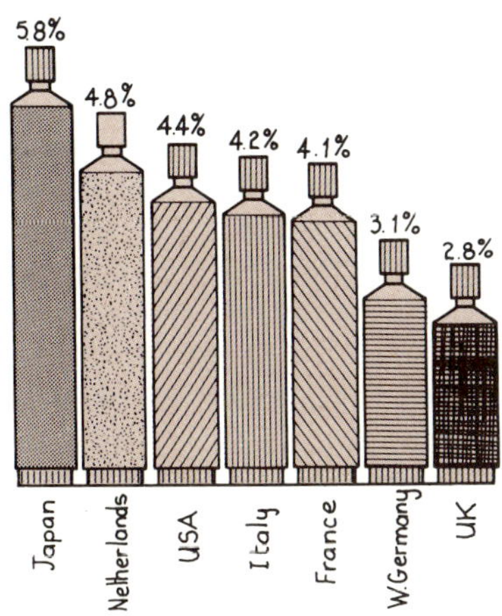

Fig. 77
Chemicals industry growth rates 1975–1985

3 After the rubbish is burnt, ash and metals are left. These are called residuals. Information about metal residuals is given in table 2.
   a   Which metal, commonly found in saucepans, is missing from the table?
   b   Which metal forms the largest part of residual composition? Suggest two household sources of this metal.
   c   Residual metal composition varies considerably. Suggest reasons why this happens.
   d   Calculate the total metal residual composition in table 2.
   e   Suggest uses for the residuals.

4 An incineration plant has five boilers. Each boiler can burn 12 tonnes of rubbish per hour.
   a   How much rubbish can the plant burn every hour?
   b   Calculate the maximum refuse that could be burnt in 24 hours.
   c   On average the plant burns 1152 tonnes of rubbish per day. What fraction of maximum usage is this? Suggest a reason for this.

5 Answer this question in groups of four students. Imagine you work for the Local Authority. It is considering building a solid waste incineration plant. You have been asked to discuss and produce:

● a list of advantages and disadvantages of such a plant

● suggestions for finding a suitable site.

Draw a poster to summarise your group's ideas. Give a talk to your class on the subject.

More rubbish

1 Fig. 76 compares the average industrial growth rates of seven different countries.
   a   Which country's industry has grown the fastest?
   b   Which country's industry has grown the least?
   c   Calculate the average industrial growth rate of the seven countries shown.
   d   On which continents are the countries situated?

2 Average growth rates of the chemicals industry in seven countries are shown in fig. 77.
   a   List four products of the chemicals industry.
   b   Which country has the slowest chemicals industry growth rate?
   c   How does the growth of the Netherlands chemical industry compare to its overall industrial growth?
   d   Find the average chemicals industry growth rate of the seven countries. How does this compare to their average overall industrial growth rate?

Answers

Sunshine
1 (d) 70%
4 (a) 6000 K   (b) 5727 °C
   (c) 697 600 km   (d) 1402.5 kg/m^3

What's for breakfast?
2 (e) 8 g
4 (b) 1472.5 kJ
   (c) wholewheat 450 kJ   bran 405 kJ

Feeding baby
1 (c) 35 g
2 (c) 21%
3 (c) 3100 kJ

Peat bogs
3 (c) 3 000 000 hectares   (d) 30 000 km^2
4 (c) 12 600 t

Microwave cooking
3 (a) 21.4%   28.6%   35.7%
      42.9%   57.1%   71.4%
   (b) 2.92 A   (c) 5 A

Bagpipes and all that jazz
3 (d) 682 Hz
4 (b) d/c 1.125   e/d 1.111   f/e 1.066

Cool customers
3 (a) can A 45 °C   can B 27 °C   can C 20 °C
4 (b) 53%

Important connections
2 (c) 1.9 A
3 (a) 3.6 Ω   4.3 Ω   5.0 Ω   5.6 Ω

Power to the people
Switch on
1 (c) 7.0 V

Cal Horrific
1 (c) 2000 kJ
3 (b) 500 kJ
4 (b) 10 000 kJ   (c) 3 years
5 (a) 11%   (b) 21.5%

Rocket sled
4 (c) 5 s
   (d) (i) 700 m   (ii) 2800 m
       (iii) 210 m   (iv) 3710 m
   (e) 56 m/s^2 (f) 56 000 N

Flying
4 (c) 3.64 m/s

Hair raising
2 (c) 2 mm/week   (d) 0.286 mm/d
3 (b) 0.75 mm/week   (c) 0.107 mm/d
4 (a) 129 750
7 (a) 148.056 mm/year   0.406 mm/d
   (b) 21.481 mm/year   0.059 mm/d

Video fission
2 (c) 7360
4 (b) 8 min   (c) 31 min   (d) 23 min

Slick experiment
More toil
1 (b) 5%   (d) 15 kg

Soap opera
Brainwashers
1 195 000 000 kg

Waste matters
3 (d) 32.92%
4 (a) 60 t   (b) 1440 t   (c) 4/5
More rubbish
1 (c) 2.59%
2 (d) 4.17%